YOUNG ENGINEER'S GUIDE

Steam Engines, 1899

J. V. Rohan

Foreword by Dan Fink

Reprinted by Buckville Publications, LLC
January 2017

YOUNG ENGINEER'S GUIDE
by J. V. Rohan

Originally published by J. V. Rohan,
Racine, Wisconsin
Original Copyrights 1894 and 1899
by J. V. Rohan

Foreword by Dan Fink
Edited by Dan Fink

Reprinted by Buckville Publications LLC
January 2017, Masonville, Colorado

Copyright 2017, Buckville Publications LLC
All rights reserved

ISBN 978-0-9819201-2-2

LCCN 2017930077

WARNING:
Steam power is an inherently dangerous activity, and a mishap can cause serious injury, severe burns, and death. Buckville Publications LLC and the editor assume no responsibility for personal injury, property damage or other loss from actions inspired by information in this book. We are simply passing along information to you from a bygone era. Safety is your responsibility.

FOREWORD

I've been fascinated by steam power since I first saw a steam locomotive at a county fair. The smoke, the sound, the danger, the incredible power of liquid water expanding into vapor. As a young teenager, I learned there was more to steam than trains: steam tractors, stationary steam engines, even steam calliopes and carousels were common. And I learned there was an old steam-powered sawmill right up the road from our family cabin high in the Northern Colorado mountains, and a rich history of logging right here in our canyon.

I was fascinated by this and wanted to learn more. Fortunately, there was a priceless resource available for my education—the Dickerson Sisters two miles up the road, who grew up here as homesteaders, moving up from down on the plains to the family homestead in a covered wagon just after the turn of the century, right after this book was published. Alice and Helen would regale me with stories about the old days whenever I hiked or rode up the road to visit them on my bike. The homemade bread and cookies hot from their wood-fired cookstove certainly had nothing to do with my frequent visits...

The logging was all for making fence posts to float down to the arid plains, and making mineshaft props for the booming gold and silver rush down south, west of Denver, for mining towns like Central City, Blackhawk, Cripple Creek and Tincup. The Dickersons had their own steam-fired sawmill to make these, and there were dozens of sawmill sites up and down the sides of the canyon. Besides sawmills, some of these sites likely also had steam donkeys and yarders.

I enjoyed exploring these old sites, and from all the stories, it made me think about what the life of a young man at a logging camp must've been like back then, whether as a steam engineer or a whistle punk.

Up two hours before first light, haul water for the steam engine up from the creek in buckets, chop and split wood for fuel, get the fire started, and have enough steam pressure by dawn to blow the whistle to signal breakfast. Repeat the water hauling, chopping and splitting all day until dark, then start again the next morning. Make a mistake, or daydream for even a few minutes, and a steam explosion could wipe you out, along with the entire camp. Running a steam engine is a lot of responsibility.

That's why I was fascinated by this book from the moment I first picked it up and started leafing through the pages, along with the end-of chapter troubleshooting quizzes and drawings. The text is readable by a teenager (or a 1900s logger with limited education), but still contains a wealth of technical information.

I hope you enjoy this blast from the past as much as I have!

> DAN FINK
> Box Prairie, Colorado
> January, 2017

YOUNG ENGINEER'S GUIDE.

BY

J. V. ROHAN.

RACINE, WIS.

REVISED AND ENLARGED

TWENTIETH EDITION.

PRICE:

CLOTH BOUND,	- - -	$1.00.
LEATHER BOUND	- - -	1.25.

PREFACE TO REVISED EDITION.

In preparing the 20th and revised edition of this book it has been the aim of the author to make such corrections and alterations as were apparent in the previous editions to add such information and illustrations as the advance in steam engineering seemed to justify, and to give some of the questions, with answers, asked of the author since the publication of the book in 1895.

I trust the revised edition will better serve the purpose intended.

J. V. ROHAN.

April, 1899.

Copyright, 1894, by J. V. ROHAN.
Copyright, 1899, by J. V. ROHAN.
All rights reserved.

PREFACE.

Some two years ago the author commenced collecting memoranda of mechanical and practical information pertaining to the care and operation of steam engines and boilers, with a view of forming a systematic digest.

Being an employee of the J. I. Case Threshing Machine Co. for a number of years, my attention was called to the constant inquiry for a book of this description from young men mechemically inclined and those running farm engines and small steam plants.

By repeated assurance that there was great need for such a work, and by ready and valuable assistance from personal friends and experts in mechanical engineering, I have considered it advisable to publish a practical guide for young engineers.

The aim has been to place the information contained in the book in the most simple and compact form, and while it is not intended for the education of the more advanced engineers, the instructions given will be found practical in the operation of steam plants of any size. It is more especially intended for the instruction and guidance of young men learning to run engines, and those operating farm engines and small plants, whose experience has been limited.

After carefully considering the mode of presentation, it was thought best to adopt the form of a catechism, with the questions and answers so set forth as to resemble an ordinary conversation; also to illustrate and give a minute description of the construction and function of the different parts used in the building of engines and boilers.

While the greater part of the information is new, parts have been compiled from Power, Roper and other mechanical papers and books, simplified to meet the required aim, for which due acknowledgement is here given.

J. V. ROHAN.

Racine, Wis., 1895.

INDEX.

PAGE.

Ascending Hills	152
Automatic Oiler	79
Banking Fires	164
Babbitting Boxes	190
Belting	167
Blower	55
Blow-off Valve	105
Calking Flues	52
Check Valve	107
Cleaning Flues	53
Compression Grease Cup.	108
Compound Engines	191
Connecting Rod	67
Crank	68
Crank-pin	69
Cross-head	66
Cross-head Pump	91
Crossing Bridges and Culverts	158
Cylinder Cocks	*106*
Descending Hills	156
Differential Gear	115
Duties of Engineers	13
Eccentric	73
Eccentric Strap	73
Eccentric Rod	74
Ejector	97
Engine Frame	67
Engine Stalled	157
Exhaust Nozzle	55
Firing with Coal	162

YOUNG ENGINEER'S GUIDE.

Firing with Wood	161
Firing with Straw	161
Foaming	158
Friction Clutch	116
Fusible Plug	56
Gauge Cocks	106
Gearing	113
General Information	172
Governor	76
Heater	96
Heating of Journals	143
Hints to Purchasers	9
Horizontal Tubular Boiler	15
Hydraulic Boiler Test	233
Indicator	226
Injector	83
Jet Pump	97
Knocks and Pounds	139
Laying Up a Traction Engine	165
Link Reverse	69
Link	72
Locomotive Boiler	16
Low Water Alarm	57
Packing Piston and Valve Rods	145
Piston and Rod	62
Priming	159
Questions with Answers concerning Boilers	30
Questions with Answers concerning Engines and Boilers	119
Questions with Answers for Engineers applying for License	197
Return Flue Boiler	18
Reversing an Engine	130
Reverse Lever	72
Rules and Tables	238
Rules for calculating the speed of Gears and Pulleys	240
Rule for finding the heating surface in a Locomotive Boiler	243
Safety Valve	101

YOUNG ENGINEER'S GUIDE.

Setting Plain Slide Valve	146
Setting Slide Valve of Reversing Engine	149
Setting Valve Duplex Pump	152
Setting Woolf Valve	195
Shafting and Pulleys	242
Steam Cylinder	61
Steam Chest	63
Steam Engine	60
Steam Engine Indicator	226
Steam Gauge	98
Steam Pump	89
Testing Piston Rings and Valves	138
Throttle	88
Traction Engines	112
Valve	64
Vertical Boiler	22
Water Tube Boiler	24
Water Gauge	104
Whistle Signals	235
Work-shop Recipes	188
Woolf Valve Gear	74

INDEX OF ILLUSTRATIONS.

ENGINES.

The E. P. Allis Co., Milwaukee, Wis	21
J. I. Case Threshing Machine Co., Racine, Wis	26, 29
Port Huron Engine & Thresher Co., Port Huron Mich	36
Nichols & Shepard, Battle Creek, Mich	49
J. T. Case Engine Co., New Britain Conn	58
M. Rumely Co., La Porte, Ind	71
Woolf Valve Gear Co., Minneapolis, Minn	75, 192
Minneapolis Threshing Machine Co., Minneapolis, Minn	111

YOUNG ENGINEER'S GUIDE.

Buffalo Pitts Company, Buffalo, N. Y.... ....134, 137
Watertown Engine Co., Watertown, N. Y........................... 129
Frick Co., Waynesboro, Pa... 155
Armington & Sims Engine Co., Providence, R. I 170
The Ball Engine Co , Erie, Pa.. 180
A. W. Stevens Company, Marinette, Wis............................ 205

BOILERS.

S. Freeman & Sons Manufacturing Co., Racine, Wis...... 15, 16, 17, 22
The Stirling Co., Chicago, Ill.. 25

FITTINGS.

Thomas Prosser & Son, P. O. Box 2873, New York City............. 52
Frontier Manufacturing Co., Buffalo, N. Y 54
The Lunkenheimer Co , Cincinnati, O.................56, 89, 104, 106, 107
The Gardner Governor Co., Quincy, Ill 76
The Detroit Lubricator Co., Detroit, Mich 81, 82, 83
American Injector Co., Detroit, Mich........................84, 87, 97, 98
Battle Creek Steam Pump Co., Battle Creek, Mich............... 93, 94
The Ashcroft Manufacturing Co., P. O Box, 2803, N. Y. City.99, 100, 226
E. B. Kunkle & Co , Fort Wayne, Ind..............................101, 102
Chas. H. Besly & Co., Chicago, Ill........ 109

Young Engineer's Guide.

HINTS TO PURCHASERS.

In selecting an engine of whatever style, or for whatever purpose it is very important to get not only a good one, but one that is of the proper size. Do not entertain the mistaken idea that it is best to have a larger engine than is required (so that it will do its work easily), as an engine which is too large for the work required is very wasteful both of fuel and water. An engine always gives the best results when it has a fair load.

In the selection of a farm or traction engine you should look carefully to the arrangement of the driving gear, the manner in which the engine and the traction wheels are attached to the boiler, the convenient arrangement of the throttle lever, reverse lever, steering wheel, friction clutch lever, independent pump (if used) and injector for easy operation from the footboard, as the easy control of all these parts by the engineer saves much time and annoyance and in many instances may prevent accident which might prove disastrous to both life and property.

Always purchase a boiler with sufficient capacity to allow a small margin beyond its ordinary requirements. Be sure and have the boiler or boilers properly set so that the best results may be derived from the fuel burned. Many good boilers are condemned because they do not steam well on account of bad setting.

If a locomotive style of boiler, see that it has a large fire box (well stayed) and a sufficient number of flues to allow of easy firing and good combustion of the fuel without being obliged to use a forced draft.

If a return flue boiler see that the main flue is of sufficient size and of the required $5/16$ inch thickness of material; also that it has a mud drum and from four to six hand holes (the more the better) both top and bottom for the purpose of keeping the boiler free from scale and becoming mud burnt and unsafe.

Remember there is no advantage in carrying low steam pressure in boilers as it is more economical to carry high pressure rather than low. The average boiler pressure should be not less than 80 lbs. for "Simple" or from 120 to 150 lbs. for "Compound" for economy in fuel.

The purchaser must use his own discretion as to the style of engine he prefers, a horizontal or vertical, side or center crank, simple or compound, as all styles are extensively used with equally good results. It is purely a matter of preference, depending, of course, largely upon space or room available for stationary engine.

Do not make the mistake of deeming that any kind of a foundation will answer for a stationary engine. It should be built by a skillful mason in every case and hard brick or stone and cement used in its construction. The best is always the cheapest in the end.

An engine or boiler should never be put in a dark corner or damp cellar, rather place them when possible in dry well lighted rooms and so arranged that every part can be reached when necessary without trouble or delay. Walls and floors should be kept clean and a good supply of oil cans, wrenches, waste and whatever tools are needed should be kept in their proper places.

The purchaser of a traction engine should see that it has a Friction Clutch as an engine with a clutch is much more practical, convenient and safe to handle upon the road than one without a clutch. The matter of brackets, braces, gearing, traction wheels, axle, the manner in which the engine is mounted upon the boiler should be well considered as there are many kinds and styles, all of which have their merit "more or less." Good judgment should be used as to the style wanted after thoroughly studying the various kinds.

Do not make the too common mistake of thinking a cheap engineer is the man you want. The engine and boiler are important factors in the success of your business and no matter how simple and strong they may be it will pay you to put them in charge of a competent engineer who is capable of taking the proper care of them. For a small plant, or traction engine, it is not necessary to have the highest grade of ability, as there are several grades among engineers; but it is better to pay a suitable man for competent and faithful work than to pay for what may happen through the neglect or incompetency of one whose only recommendation is that he is *cheap*.

Do not be deceived by imposters claiming to be first-class engineers, who, the first thing they do, to substantiate their claims, alter the engine in some way that only deranges it. Be watchful of this and see that such men do not tamper with the valves and adjustments of the engine, which are always set properly before it leaves the factory.

DUTIES OF ENGINEERS.

The duties of an engineer are of much more importance and require a better knowledge of the operating of machinery than is generally understood. The responsibilities that rest upon him are very great; this applies to all engineers, but more especially to inexperienced men who take charge of small plants or farm engines, whose knowledge of machinery and the dangers connected with the improper handling of it, is limited. The proper management of boilers and engines is of as vital importance to prevent accident as their proper construction; as they are liable to get out of order and become unsafe unless the engineer is sufficiently informed to know what precautions should be taken under any and all circumstances that might prove disastrous.

Not only should an engineer be ever on the alert to guard against accident, but he should also be capable of keeping the engine, boiler and appliances in good condition, as the life of the machinery depends largely upon his competency and the faithful performance of his duties.

An ENGINEER:

Should be sober.

Should be industrious.

Should be careful.

Should be faithful to his charge.

Should keep his engine and its surroundings neat and clean.

Should keep his engine running smoothly without knocks or pounds.

Should learn to let "well enough" alone.

Should never attempt experiments unless he knows what he is about.

Should have a place for everything and keep everything in its place.

Should show by the quietness in running and appearance of the engine in his charge that it is properly cared for.

Should constantly endeavor to expand his mind as to the management, construction and care of boilers, engines and their appliances.

Should carry this book in his pocket for reference as it contains much valuable information and in a time of need may save much time and expense, or even prevent a catastrophe.

Boilers.

HORIZONTAL TUBULAR BOILER.

Q. How is a horizontal tubular boiler constructed?

A. It has a cylindrical shell, with heads riveted at each end, in which are placed a large number of tubes, 4 inches or less in diameter.

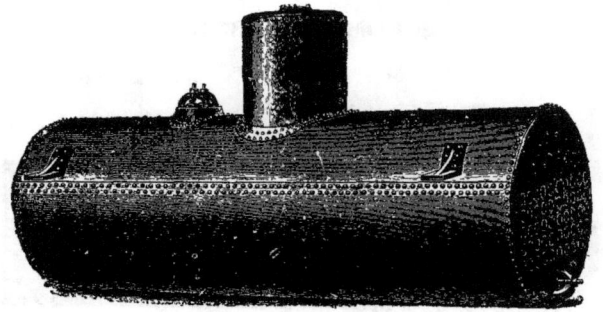

Freeman Horizontal Tubular Boiler.

It is set in brick-work, with the furnace at one end, beneath the shell. The products of combustion pass under the boiler its full length and return through the tubes or flues to an up-take at the front end. It is furnished with a man-hole beneath the flues, and hand-holes for cleaning, and generally has a steam dome.

Q. What are the advantages of a horizontal boiler?

A. It is simple in form, easy to construct, requires bracing only on the flat heads, which are sustained their greater part by the tubes, generates a large amount of steam for the space occupied, and is not difficult to keep clean with fairly pure feed water.

LOCOMOTIVE BOILER.

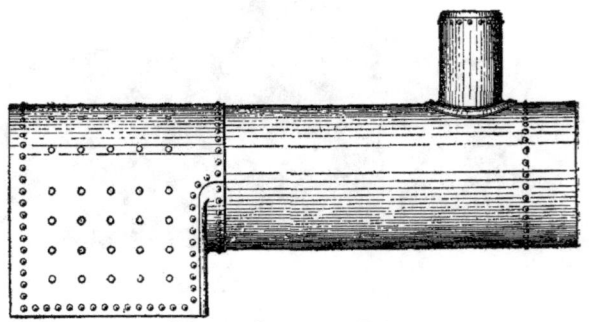

Freeman Locomotive Boiler.

Q. How is a locomotive or fire-box boiler constructed?

A. The ends of a cylindrical shell are continued straight down upon the sides, and enclosed to form a rectangular structure in its lower portion and conformed to the curvature of the cylindrical shell at the top. In the rectangular portion is secured a fire-box, separated from the sides and ends by water spaces called "water legs," and having its top which is called the "crown sheet" about the center of the cylindrical shell. An opening is formed in both sheets in rear end of the fire-box door frame. The cylindrical shell has heads riveted at both

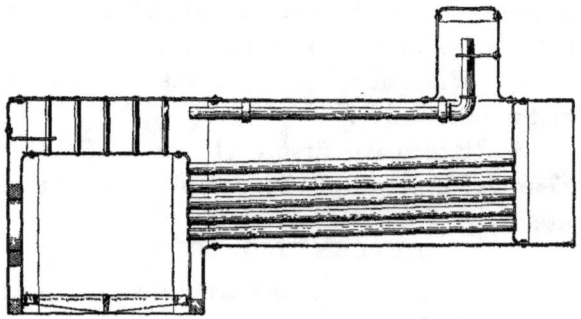

Sectional View Freeman Locomotive Boiler.

ends. These heads have numerous flues open at both ends put in. All the flat surfaces are stayed to each other at suitable intervals, and the crown sheet is stayed from the shell. The fuel is thrown in through the door at the rear, and the products of combustion are conveyed to further extremity through the tubes or flues. It is furnished with hand-holes for cleaning and a steam dome. There are two different styles of fire-boxes on Locomotive Boilers, the round bottom fire-box in which the water circulates under the grates, and the square open bottom fire-box. Both kinds are used extensively.

Q. What advantages have the locomotive or fire-box boiler?

A. It is entirely self-contained, generates steam very rapidly, is economical in space, and needs no elaborate foundation.

Q. What disadvantages has the locomotive or fire-box boiler?

A. Expensive first cost, and difficulty in cleaning, especially where impure feed water is used.

RETURN FLUE BOILER.

Q. Describe the construction of a Return Flue Boiler?

A. It has a cylindrical shell, with heads riveted at each end, in which are placed a large main flue, and a number of small flues or tubes, open at both ends. The top row of flues is placed below the water line. One end of the main flue is used for the fire box, into which the fuel is thrown through door at back end and the products of combustion pass forward through this main flue to an ample smoke box in front end, and return through the smaller flues or tubes to smoke box at rear end, which is connected to the smoke stack. They are generally supplied with steam dome and mud drum, and are used extensively in the construction of traction

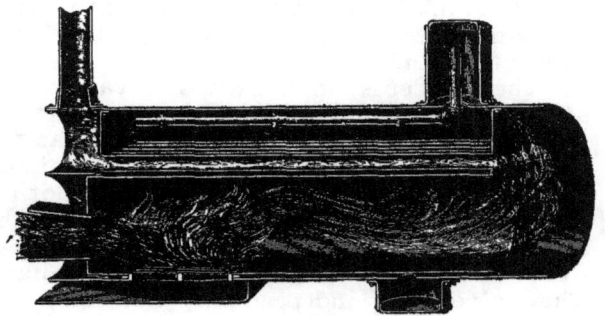

Sectional View Return Flue Boiler.

engines where straw is used for fuel. They are also furnished with several hand-holes placed in proper places for cleaning the boiler.

Q. What are the advantages of this style of boiler?

A. Large heating surface, easily repaired and cleaned, simplicity of construction and compactness.

REYNOLDS CORLISS CONDENSING AND NON-CONDENSING ENGINE.

This engine has a massive, peculiarly constructed frame, being cast in two parts. The forward part contains the main pillow block bearing, and the part in which the cross-head runs is cylindrical in shape, and has bored Guides, and large lateral openings.

The crank is a large disc plate, and the large heavy fly wheel serves the double purpose of a drive pulley and balance wheel.

The cylinder is supplied with four valves, two at the top and two at the bottom, and directly upon the bore of the cylinder. The two at the top are the steam valves, the two at the bottom are exhaust valves, and receive their motion from a single eccentric acting through the medium of a wrist plate or vibrating disc, from which the valve connections radiate. The valve being independently adjusted, the commencement, extent and rapidity of the movement of each can be most accurately arranged. The steam valves are controlled by the Governor, which, being very sensitive to the variation of load, allows just the required amount of steam to enter the cylinder to keep up the uniform speed. The exhaust valves being at the lower ends of the cylinder, at the clearance space, the water of condensation is allowed to escape in the most thorough manner, without the use of cylinder cocks or other devices.

This style of engine is intended for all purposes, but is especially adapted to heavy and continuous work, and where the work is of an intermittent character.

Reynolds Corliss "1890" Engine—Front View.

VERTICAL BOILER.

Sectional View Freeman Vertical Boiler.

Q. How is a Vertical Tubular Boiler generally constructed?

A. A cylindrical fire box set into the lower part of a vertical cylindrical shell, the space between forming an annular "water leg." An opening is formed in both sheets for the fire door. The top of the fire box serves as a flue sheet for numerous tubes or flues which extend through the closed top of the outside shell, and through which the products of combustion pass to the smoke stack. The upper portions of the tubes are surrounded by steam.

Where this style of boiler is made for marine purposes, the upper part of the tubes is submerged, and is called a submerged-flue boiler.

Q. What advantages has the vertical type of boiler?

A. Minimum floor space, portability, low cost of setting, and a wide allowable variation in the water level.

Q. What disadvantages has this type?

A. Liability to leakage in the exposed upper ends of flues where they are not submerged, deposits from impure water in the "water leg," in small sizes insufficient heating surface, though the latter fault can be corrected by making the boiler very tall. Some of the very large vertical boilers are remarkably efficient.

WATER TUBE BOILER.

Q. How is a Water Tube Boiler constructed?

A. The specific difference between a Return Tubular, or Fire Tube Boiler, and a Water Tube Boiler lies in the fact that in the former the prod-

Stirling Water Tube Boiler.

ucts of combustion pass through the tubes on their way to the stack, whilst in the latter they pass around the tubes. There are a great many different forms of Water Tube Boilers, and the limits of this publication render a full description of all the types impossible. A few words in relation to the Stirling, however, may be in order, as that boiler is one of the most prominent on the market, and offers many points of superiority and excellence. As appears from the cut opposite, the boiler comprises three upper or steam drums, and one lower or mud drum, all connected by means of wrought iron tubes which are expanded in each drum. The feed water enters the rear upper drum and in its descent to the mud drum becomes heated by the escaping gases to a sufficiently high degree to cause the precipitation into the mud drum of all the solid or scale bearing matter that it may contain. This sedimentary matter may, in turn, be blown off as rapidly as it accumulates. The front banks of tubes are, therefore, filled with chemically pure water, and all danger of scale reduced to a minimum. Cast iron and stay bolts, which are such distinct disadvantages in other types, have been entirely overcome in the Stirling. The boiler is built entirely of steel and requires no stay bolts and has no flat surfaces. It is simple, economical in operation and easy to clean, each drum being provided with a man hole by which access to the interior of every tube can be had.

A large amount of power can be installed in a small space, rendering the boiler particularly well adapted to large plants where large installations of power are required.

J. I. Case Traction Engine.

J. I. CASE TRACTION ENGINE.

This engine is known as the Side Crank Spring Mounted "Compound" Traction and is of the side gear design with engine mounted upon the side of the boiler. The engine frame is supported by a triangular bracket, bolted to side of boiler, and a pillow block bearing which is bolted to and held in position by the extension of the side plate of boiler at the fire box end.

The frame is of a girder pattern, cast in one piece, with bored guides for cross head and has large lateral openings. It forms the back cylinder head at one end and contains the pillow block bearing, for crank shaft, at the other

The pillow block bearing for opposite end of crank shaft is held in position and bolted to the extension of side plate of fire box.

The cylinder is the Woolf, Tandem Cylinder, Compound type; it is overhanging and self-lining.

It has the locomotive type of boiler with steam dome and an open bottom fire box, covered with an ash pan.

The side plates of fire box extend up even with top of the shell. These extensions form the foundations for the crank shaft bearings and to which the upper radius link brackets are bolted.

The main axle and countershaft have bearings in two cannon bearings which are held in position, at rear of boiler, by six radius links and two distance bars. The lower cannon bearing supports the spiral springs upon which the boiler rests.

The traction wheels are of the wrought rim steel spoke type with high mud cleats riveted diagonally across the entire width of tire.

The front axle is similar to a spring girder drawn together at the ends and is supplied with flanges to which the steering chains are attached. The front bolster is bolted to boiler plate bracket riveted to the shell of the boiler.

A cast iron spark arresting stack is used.

The differential gear is supplied with spiral spring as is also the platform. The water tank is placed under the platform.

The steering wheel and band wheel are on the right hand side of engine.

The engine has a Friction Clutch, Woolf Single Eccentric Reverse Gear, Long Heater, Independent Pump and Injector at rear, tool box and all fittings and attachments necessary to complete it for the purpose intended and it will be perfectly safe with proper handling.

The boiler is constructed to burn either coal or wood. It can also be made to burn straw efficiently by placing a fire brick arch of special design within the fire box.

This engine is of new design with many radical changes which will be seen by examining the illustrations.

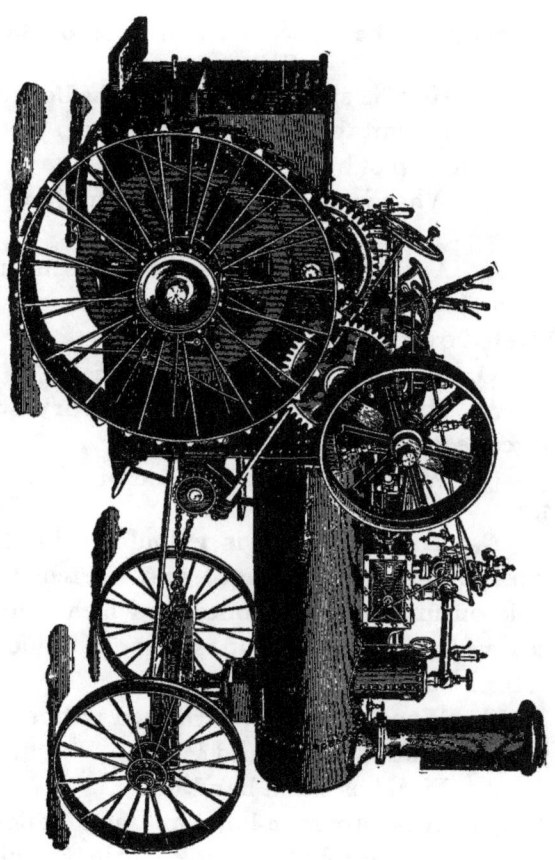

J. I. Case Traction Engine.

QUESTIONS WITH ANSWERS,

Concerning the Operation and Care of Steam Boilers.

Q. How is steam taken from the boiler?

A. By suitable piping leading from a steam dome on top of boiler.

Q. What is a steam dome and how is it made?

A. A steam dome is cylindrical in shape, is made usually of boiler plate flanged and riveted over a hole on top of boiler.

Q. Of what use is a steam dome?

A. Its use is to afford space for dry steam to collect.

Q. What is a mud drum, and of what use is it?

A. The mud drum is cylindrical in shape, made of boiler iron flanged and riveted over a hole on under side of boiler, in which mud and sediment may collect, and is of great value on a return flue boiler.

Q. What are boilers furnished with so that they may be easily cleaned?

A. Man-holes and hand-holes.

Q. What are man-holes and hand-holes?

A. The man-hole is a hole cut in boiler large

enough to admit a man and is covered by a portable plate which can be fastened absolutely tight. Hand-holes are small holes cut in boiler in convenient places into which a hose can be placed and the boiler washed out and the mud and scale removed. Hand-holes are covered with portable plates which can be fastened absolutely tight.

Q. How can a boiler be protected from the cold?

A. By a non-conducting jacket which keeps off the cold, retains the high temperature of the boiler and prevents the radiation of heat.

Q. What materials are used for jacketing a boiler?

A. Plaster, wood, hair, rags, felt, paper and asbestos.

Q. How are they applied?

A. Wood is put on in long strips close together like barrel staves, fastened with hoops and usually covered with sheet iron. The other materials are put close to the boiler and held in place by sheet iron or canvas. They are sometimes put on with an intervening air space between them and the boiler.

Q. What is the use of air space?

A. It protects the material from being burned or otherwise injured by the heat.

Q. Is there no radiation through these coverings?

A. Yes, but the loss is very slight, as the temperature of covering should never rise above what just seems warm to the hand.

Q. How should you feed water to a boiler?

A. Continuously during the whole time that steam is being used.

Q. Will a steam pump feed continuously?

A. Yes, by running the pump faster or slower according to the amount of water required.

Q. Why is a continuous feed preferable?

A. Because it maintains the water in the boiler at a uniform level and gives the most perfect action.

Q. Should precaution be taken in choice of water used in the boiler?

A. Yes. Always use water that is as clear and free from foreign matter as can be procured, rain water preferred.

Q. What is the result of using impure water?

A. It will form a scale upon the flues and plates on the inside of boiler.

Q. What harm does scale do?

A. In the first place scale is a non-conductor and prevents the heat of furnace from producing its best effects upon the water and in the second place it allows the plates and flues to become over heated and burn.

Q. How can you prevent the formation of scale?

A. There are numerous compounds upon the market some of which are known to be very reliable, but for ordinary purposes sal soda dissolved in the feed water answers very well.

Q. What precaution should be taken in the use of sal soda?

A. Great care should be taken that too much is not used at a time. If too much is used a great deal of trouble will be caused by the water in the boiler foaming.

Q. How often should a boiler be cleaned?

A. It depends entirely upon the condition of the feed water used and the amount of service exerted from it. It may vary from once or twice a week to once in two or three months, or even longer.

Q. Does a boiler only require the regular cleaning?

A. No, it should be blown off three or four times a day by the surface blow-off.

Q. Should the surface blow-off be left open any length of time?

A. No, only a few seconds at a time, say from fifteen seconds to a minute, even longer on larger boilers, but the engineer must use his own judgement in this matter.

Q. What does the surface blow-off do?

A. It blows out all the impurities that arise in the form of scum on the surface of the water, thus lessening the amount of scale formation.

Q. How should a boiler be cleaned?

A. By blowing the water out at a low pressure of steam and after cooling off wash out and scrape the inside, removing all scale and sediment.

Q. How do you blow off your boiler?

A. By means of a blow-off valve situated at the bottom part of the boiler.

Q. When should a boiler be blown off?

A. When the steam pressure entirely disappears and the water is at boiling point, if boiler is set in brick work.

Q. Why not blow off under a full head of steam?

A. Because when blown off under pressure there is heat enough remaining in the shell and flues to bake the scale upon the interior, thus rendering it exceedingly difficult to remove.

Q. In what condition is the scale after blowing off at low pressure.

A. Some may be baked hard and attached to the flues and shell, but the greater part will be soft and slushy so that it can be easily removed.

Q. How is this slush removed from the boiler?

A. If a fire box or return flue boiler, all the hand-hole plates should be removed and as much of the slush as possible raked out; then a hose is inserted and a stream of water forced in which will carry the remainder out. The hose should be placed in the top holes first.

Q. Is it a good idea after blowing off a boiler to fill it with water again without delay?

A. No. Because the boiler is hot, and if cold water is put in before it is thoroughly cooled off the boiler will be subjected to severe strains caused by the sudden contraction of the metal that is expanded by the heat, which will injure it to a greater or less degree.

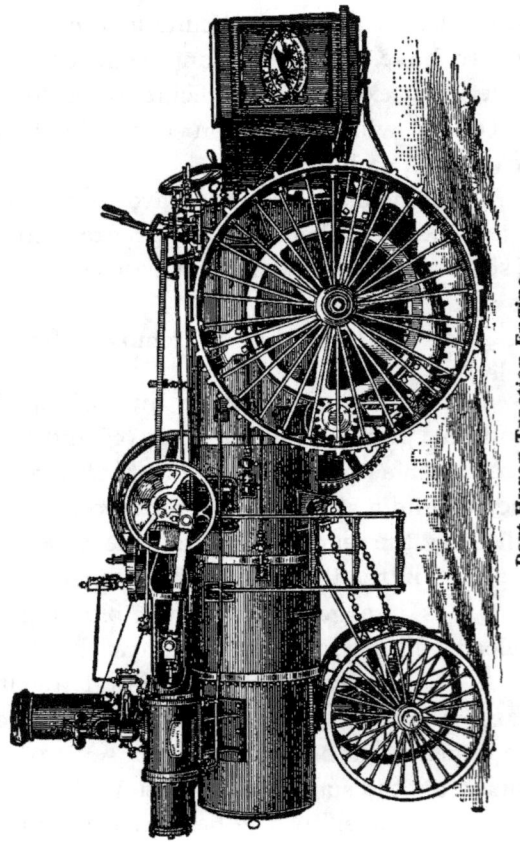

Port Huron Traction Engine.

PORT HURON TRACTION ENGINE.

The cut opposite shows a Compound, Side Crank Traction Engine of the Side Gear type, with the engine placed well forward upon the boiler to allow of short steam and exhaust pipes.

The engine frame is of the girder pattern, has bored guides with large lateral openings and contains the pillow block bearings for crank shaft. The frame is held in position at crank end by a bracket bolted to side of boiler, the front end is bolted to, and forms the back head for low pressure cylinder, which is secured to boiler by bracket bolted to smoke box end.

The Woolf Tandem Compound Cylinders and Grime Single Eccentric Valve Gear are used in its construction. The band wheel and steering wheel are on right hand side of engine.

The boiler is of the locomotive round bottom fire box type, with dome a little forward of the center, is nicely lagged and jacketed and has a long smoke box. It is mounted upon the traction wheels with axle arms attached to side of fire box.

The traction wheels are made practically in one piece having the spokes cast in both hub and rim, and high mud cleats cast on rim.

This engine is symmetrical in its design and has tool and fuel boxes on platform, step on side, and all fittings and attachments necessary on a Traction Engine to make it safe, economical and easy to handle.

Q. When should a boiler be filled after being blown off and cleaned?

A. A boiler should not be filled under any circumstances until it is about the same temperature as the water used to fill it.

Q. How are the hand-hole plates put back in position?

A. The plate that covers the hole is put on the inside of the boiler and held in position against the plate by a bolt attached to it and passing out through a yoke which straddles the hole upon the outside and serves as a brace against which the bolt acts.

Q. How is a leak prevented?

A. By placing packing between the hand-hole plate and the boiler plate.

Q. What is the best kind of packing to use for this purpose?

A. Two or three ply sheet rubber is the best, cut in the form of a ring to fit the bearing surface of the plate.

Q. Can any other material be used?

A. Yes, hemp or cotton packing. When this is used it should be pulled out in fine shreds and thoroughly oiled before putting in position. Use as little as possible.

Q. Do the metals need any preparation?

A. Yes, all the old packing that may have burned on the metal should be thoroughly scraped off, also the scale should be removed from the vicinity so that the packing will have a smooth and even surface to bear against.

Q. How are man-hole plates put in position?

A. In the same manner as the hand-hole plates.

Q. What is required of an engineer or fireman in the care of a boiler?

A. He should watch carefully all the parts that are exposed to any steam and see that they do not become unduly weakened by corrosion or accident. All the working parts and fittings should be examined daily and be repaired or replaced as soon as they show signs of wear or weakness. The steam gauge and safety valve should receive constant care and both should be tested frequently, the one by the other, and the steam gauge by a standard in order that it may be known to be in perfect order. When scale forming water is used the feed pipe should be frequently uncoupled and examined and all sediment removed. The check valve should be examined frequently to see that it seats properly so that water cannot leak from boiler in this way and the utmost care should be taken in regard to the consumption of fuel.

Q. In case of accident how should an engineer conduct himself?

A. With the utmost coolness.

Q. If the water gauge glass breaks, what should be done?

A. The upper and lower gauge valves should be closed immediately.

Q. Can a new glass be put in at once?

A. No, because a new glass is cold, and if put in position and steam turned on, the sudden heat and expansion would be apt to crack it.

Q. When can a new glass be put in?

A. After the boiler has been cooled off.

Q. What is to be done in the meantime?

A. The boiler must be run by the use of the gauge cocks alone.

Q. What is to be done if the gauge cocks leak?

A. If the leak is in the seat, that part should be taken out and re-ground and refitted. This should be done at once.

Q. What harm is done by a leaky gauge cock?

A. It allows the water to run down over the face of the boiler, which tends to corrode it.

Q. When the leak is where the gauge cock is screwed into the boiler, what is to be done?

A. As soon as the boiler is cooled down examine and see if the gauge cock can be screwed up another turn. If so try that, then if the leak is not stopped the gauge cock must be taken out and a new one put in its place, or the thread of the old one so repaired that there will be no leak.

Q. Why not screw up the leaky gauge cock when the boiler is under pressure?

A. Because there is great danger of breaking the cock, thereby placing the engineer or fireman in great peril.

Q. What should be done in case a gauge cock is accidentally broken off?

A. Open the furnace door and if possible partially bank the fire, close the damper and allow the water to blow out at the hole until steam alone comes out. In the meantime get a piece of soft pine six or seven feet long and whittle down one end until it will about fit and jam it into the hole. Work it around until the jet of steam is stopped. Fasten the stick in temporarily and stop the engine if not already done. It will now depend on the condition of the break and the position of the surrounding parts as to the means to be employed. The stick should be cut off short and firmly driven into the hole and braced or tied securely. The engineer or fireman must use his own ingenuity for this work.

Q. Can a boiler be worked in this condition?

A. Yes, by the use of the gauge glass to determine the level of water.

Q. Should a boiler be run in this condition continually?

A. No. A new gauge cock should be supplied as soon as possible.

Q. When a gauge cock becomes stopped up what should be done?

A. After steam is down, the front or outer part may be taken off and a stiff wire run into it to open the clogged tube.

Q. Is it simply necessary to get the wire through?

A. No, the wire should be worked back and forth until all the deposit or scale is thoroughly cleaned out.

Q. In case the steam gauge gets out of order what should be done?

A. There should always be an extra gauge on hand that may be put to use. If there is no extra steam gauge, the engine should be shut down until the gauge can be repaired.

Q. Why not continue running by using the safety valve?

A. Because it is very dangerous and should never be attempted.

Q. How much variation from the actual pressure can be allowed on steam gauge before it is repaired?

A. None. As soon as suspected of being even slightly out of order it should be repaired.

Q. In case the pump does not work what should be done?

A. Supply the boiler by the injector.

Q. What is to be done where there is no injector?

A. First, care should be taken that the water in the boiler does not fall below the second gauge cock or out of sight in the gauge glass, then stop the engine and bank the fire. When this is done, examine the packing around the plunger to see that it does not leak air, then examine the valves of the pump to ascertain whether they are worn and leak. If this be the case they must be reseated at once. If the valves are all right, work the pump and open the side valve in the delivery pipe to see if the pump draws water. If no water appears, the trouble is probably in the suction pipe.

Q. How can this be remedied?

A. First examine the strainer at end of suction pipe or hose to see if it is stopped up; if it is, clean it out and try the pump. If it works, the difficulty is remedied. If the strainer is clean, examine the pipe or hose from end to end to see that it is perfectly air tight; if not, it should be made so.

Q. If the delivery pipe is choked, how can it be cleaned?

A. Close the globe valve next to boiler, and then examine the check valve to see if it is all right. If it is choked or filled with sediment, take out the valve, clean the shell and re-seat the valve; if the check is all right, disconnect the pipe and clean out if necessary.

Q. What if this pipe and check valve are all right?

A. Let the boiler cool off, then blow off the water, disconnect the pipe between check and boiler, where the difficulty will probably be found, and clean thoroughly.

Q. How can check valve and delivery pipe be choked with water that has already passed through the injector or valves of the pump?

A. The water may contain quantities of lime which are deposited from the heated water upon the interior of the pipe, which will thus be gradually decreased in size until the hole is too small to answer the purpose.

Q. When the communication between the water gauge and boiler is interrupted, what should be done?

A. The glass should be blown out frequently by opening the drip cock at the bottom, then shut the upper valve, allowing the water to blow through the lower valve until the water runs free and clear. Then shut the lower valve and open the upper one and blow through in like manner.

Q. In case of low water what should be done?

A. Cover the fire quickly with fresh coal or damp ashes, close the lower draft door, and allow the furnace to cool. Never dash water into the furnace to check the fire, it is dangerous.

Q. Why not draw or dump the fire?

A. Because it would result momentarily in stirring up an intense heat, cooling can be effected more rapidly by covering the fire and checking the draft.

Q. Should the pop or safety valve be opened?

A. No. Never let more steam out of the boiler in this condition than can be avoided.

Q. Should the engine be stopped or the throttle valve be closed?

A. No. A sudden stoppage of the outflow of steam will cause the water level to fall. The first thing to be looked after is to subdue the heat which is the source from which trouble may arise.

Q. Should the feed water supply be turned on?

A. No. *Leave it alone.* Should the pump or injector be running, the water level will be recovered gradually as the boiler cools down. If the feed is not on, the sudden admittance of water on the overheated surfaces will cause a disaster. The feed should not be turned on until sufficient time has been allowed to avert such danger.

Q. Are there any appliances by which to guard against accident from low water?

A. Yes. Alarms to call attention by blowing a whistle or ringing a bell when the water is below a certain level. Also fusible or safety plug placed in the heating surface of boiler most liable to be overheated from lack of water.

Q. Of what use is the safety valve?

A. To prevent the accumulation of pressure above a given point.

Q. Should water be left in the boiler when not in use?

A. No. It is better to draw out all the water and properly clean the boiler before leaving.

Q. What should be done in case a grate bar breaks and drops out of place?

A. If no other bar is at hand, it might be repaired with a heavy stick of wood.

Q. How can this be done?

A. By cutting the stick in such a shape as to fit the space made by the broken bar, then cover with ashes before the fire spreads over it.

Q. Will the stick burn out?

A. Yes, but it will last for several hours.

NICHOLS & SHEPARD TRACTION ENGINE.

This engine is also of the Side Crank, Side Gear pattern, the engine is mounted upon the side of the boiler, upon a long heater, which is securely bolted to the boiler by three brackets. It is of the Locomotive Guide pattern, and has a cross head pump. The cylinder rests its full length upon the heater and is lagged. It has the link reverse gear and plain slide valve, Friction Clutch, Injector, Automatic Sight Feed Lubricator, Governor, Extension Front and Straight Stack. The hand steering wheel is on the opposite side from the band wheel.

The boiler is of the round bottom fire box style with double riveted seams, and has a dome in the center, and is mounted upon the wheels in the rear with a wrought iron axle which passes around underneath the boiler and is held in place by brackets attached to the side of the boiler.

These brackets contain springs.

The traction wheels are of the cast iron rim style, with wrought iron spokes cast in both rim and hub and mud cleats are cast on rim. The foot board is furnished with a water tank and tool box and all necessary fittings and attachments are supplied to make a complete traction engine.

YOUNG ENGINEER'S GUIDE. 49

Nichols & Shepard Traction Engine.

Q. What harm would result from firing for a short time without the bar?

A. None to speak of to furnace or grate bars, but the quantity of air admitted to the fire box would make it exceedingly hard to keep up steam and the hole thus made would cause a great loss of fuel.

Q. What should be done if a bar in a rocker grate should fall out?

A. Take a piece of flat wrought iron and cut it to fit the bearings, this will do for some time and will not interfere with the rocking of the bars. Or, take a heavy piece of plank, covering the opening completely, and cover the plank carefully with ashes, surrounding it on all sides to protect it as much as possible from the fire.

Q. Will not this latter prevent the rocking of the grate?

A. Yes, and it can only be cleaned by raking out from underneath.

Q. How should such difficulties be avoided?

A. A good engineer will always have on hand at least two or three extra grate bars.

Q. Should a boiler be forced beyond its normal capacity?

A. Never force a boiler beyond its normal capacity, as such excessive firing distorts the fire sheets and results in leaks and fractures.

Q. Should intense fires be started in or under boilers?

A. Never build an intense fire in or under a boiler until the shell is well heated. Hot fires in or under cold boilers hurry their destruction.

CAUTION.

Never blow out the boiler under high steam pressure or fill it again with cold water when the boiler is hot, as either one of these is likely to fracture the transverse riveting and is dangerous.

Do not feed the water to the boiler irregularly. The slower the water goes through the heater the more heat it takes up. To fill the boiler to three gauges and then shut off the feed until the water level is again down, the exhaust steam, after it has heated the water standing in heater, passes off without leaving any of its heat, and then turning on the feed water again much faster than needed, the water has not time to take up so much heat as if the feed were slow and regular. Much bad effect on the boiler is due to the difference in temperature at the time when the feed is off and on.

Never calk a boiler under steam pressure unless you are tired of life.

CALKING FLUES.

Q. In case the boiler flues become leaky, can they be tightened?

A. Yes, by the use of a tool called an "expander," which is generally kept in stock by the boiler manufacturer.

Prosser's Spring Expander.

Q. Can an unskilled person expand and tighten the flues of his boiler without the aid of an expert or boiler maker?

A. Yes, if he is careful and follows these instructions, viz:

First clean the ends of the flues and flue sheet of all dirt, soot and cinders, and place the expander within the leaky flue, being careful to have the shoulder of the tool well against the head and end of flue. Now, with a light hammer drive in the taper expanding pin and after two or three blows of the hammer, jar the pin out, turn the expander a little and drive in pin as before, removing the pin and turning expander again until a full turn has been made.

Great care must be taken in expanding flues not to expand them so hard as to stretch or enlarge the hole containing flue in the flue sheet, thereby loosening the adjoining flue. After all the loose flues have been carefully expanded, take the beading tool and place the long or guide end within the flue, then with the aid of a light hammer the ends of flues can be gradually beaded or calked against the flue sheet, rendering them perfectly tight. With a little practice, a careful man can do a neat job of calking, thereby avoiding loss of time and expense in being obliged to call a boiler maker. An expander and calking tools should be among the tools of every engineer, as many little leaks that may occur in a steam boiler, although they may not be dangerous, give it a bad appearance and should be calked and stopped by the engineer.

CLEANING FLUES.

Q. How are boiler flues cleaned?

A. There are two ways of cleaning flues, viz: with a steam blower and a scraper. The latter is more commonly used and when properly applied does its work most efficiently.

The scraper is screwed on the end of a rod of sufficient length to allow it to pass through the flue, and when cleaning, the scraper should be passed forward and backward through the flue or least two or three times, to insure all the soot and ashes being removed.

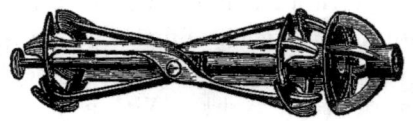

Wilson Pat. Flue Scraper.

The cleaning of flues should be done systematically, as often as required to keep them clean, as clean flues not only add greatly to the steaming capacity of the boiler, but make a great difference in the amount of fuel used.

If the flues are allowed to become covered inside with soot and ashes, the formation of which becomes a non-conductor for heat, the product of combustion passes through the flues without leaving more than one-half as much heat as it would otherwise leave if the flues were clean.

BLOWER.

The Blower consists of a small pipe attached to the steam dome or top of the boiler with a globe or angle valve, and is situated near the stack. This pipe enters the stack just above the boiler, the end being bent up toward the top of stack and reduced to a very small opening. When steam is turned on through this pipe, it displaces the air in the stack, causes a partial vacuum in the smoke box, and the air rushes through the grates, fuel and flues to replace that which is blown out by the blower, and the draught can be increased as much as desired.

EXHAUST NOZZLE.

The Exhaust Nozzle, as generally constructed, is an elbow attached to the end of the exhaust pipe in the smoke box or smoke stack of the boiler, the end of which points upward, with the opening reduced so that the exhaust steam will be forced up the stack, and thereby produce the same effect as a blower.

The opening in the exhaust nozzle should never be made so small as to check the exhaust steam to any great degree, and cause back pressure in the cylinder, as the power of the engine would be diminished. The opening should be as large as possible, and still produce sufficient draught to keep the required steam pressure.

FUSIBLE PLUG.

Lunkenheimer Fusible Plug.

A **Fusible Plug** is a short brass bolt which has a hole running through its center, filled with a metal that melts at a low temperature. This plug is screwed into the crown sheet directly over the fire, and as long as it is covered with water the metal will not melt and run out; but should the water become low, exposing the crown sheet to the intense heat of the fire, the metal will run out, and the steam rushing through the hole puts out the fire and many times saves the crown sheet from injury.

Q. In what condition would the plug become useless and of no value?

A. By allowing it to become covered on the inner end with scale and sediment. It should be unscrewed and occasionally examined, at least two or three times during the season, and all scale and dirt removed from the end of the plug before replacing. Examine the crown sheet to see that no scale has formed over the hole to prevent the water from reaching the plug.

Q. How can a plug that has melted out be refilled?

A. Unscrew the plug from crown sheet and cap one end with clay, then melt the lead or babbitt metal in a shovel, spoon, or even a piece of bent sheet iron, and refill the plug. Now, with a light hammer, close the ends tight, and screw the plug into the crown sheet.

LOW WATER ALARM.

A low water alarm is an instrument attached to a boiler, and so arranged and constructed that when the water in the boiler gets to a certain level, whereby it is becoming dangerously low, the alarm is given by the blowing of a whistle or ringing of a bell.

J. T. CASE AUTOMATIC HIGH SPEED ENGINE.

This style of Automatic Engines combines simplicity, compactness, direct action, lightness of moving parts, automatic lubrication, and perfect regulation.

J. T. Case Automatic Pedestal Engine.

It is made in three main varieties: The Pedestal Engine, the Bracket Engine, and the Hanger Engine. The illustration represents the Pedestal type, the upright frame of which is cast in one piece, and encloses and protects the principal moving parts, its lower part being a reservoir for oil, into which the crank pin dips at every revolution, affording a simple and efficient means of lubrication.

The piston is connected directly with the crank shaft, thus doing away with the cross head, wrist pin and guides. The piston being thus connected at one end to the crank pin, it travels back and forth at its other extremity through the bore of the cylinder. The latter by reason of its shape is free to turn in its casing, and is therefore rocked by the vibrating piston rod through an arc sufficient to open and close the steam and exhaust ports on its face.

The cut-off valve is of the plug type, and receives its motion from the shaft cut-off governor, attached to the balance wheel.

The center crank shaft runs in two large bearings which are bolted securely to the side of the frame. Access to the inside of the frame can be had by taking off the plates from either side.

The J. T. Case High Speed Engines range in size from 2½ to 25 horse power, the speed of which ranges from 900 down to 550 revolutions per minute, and can be used in any capacity.

STEAM ENGINES.

All styles of engines both large and small should receive proper attention. All the vibrating and moving parts should be kept well oiled and free from grit and dirt. If this is neglected, the friction of the moving parts will soon wear away the metal and induce pounding and cause what is called "lost motion," which detracts greatly from the power of the engine, and if allowed to run in this condition will soon necessitate large expense for repairs and shorten the life of the engine.

Tighten all the boxes as they wear, being careful not to get them too tight. Keep the piston rod and valve rod well packed with a good quality of soft packing. Keep the valve or valves set properly to give the required amount of lap and lead and an equal cut-off at the end of each stroke whether working in full gear or notched up.

Keep the cross-head shoes fitted properly in the guides, being careful to keep the piston rod in line. If the above instructions are followed, your engine will run smoothly and do good service.

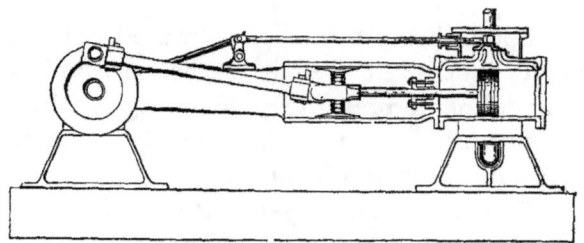

Sectional View of Simple Engine,

Showing Cylinder, Steam Chest, Plain Slide Valve, Steam and Exhaust Ports, Piston and Rod, Engine Frame, Crosshead, Connecting Rod, Crank Disc, and Rocker Arm and Rod for Operating Valve.

STEAM CYLINDER.

The Steam Cylinder is that part of an engine in which the piston travels; it also contains the steam and exhaust ports and is one of the most expensive, as well as essential, parts of an engine. The cylinder should be made of the best quality of cast iron, and the greatest care taken in boring it perfectly true and round. It should be counter-bored at each end to allow the piston in its travel to overlap at the end of each stroke. Without the counter-bore, a shoulder would be formed at both ends of the cylinder as it became worn by the piston rings, which in time would cause a knock or pound at the end of every stroke; the only remedy being to have the cylinder re-bored.

Special attention should be paid to keeping the cylinder well oiled with the best quality of cylinder oil to prevent it from being cut by the piston rings. If allowed to run dry and cut, it will cause no end of trouble.

The size of cylinder is not always the measure of the power of the engine. The power depends upon the heating surface of the boiler and steam pressure; as the piston speed can always be increased, by running the engine faster, until the maximum evaporating capacity of the boiler is reached.

PISTON AND ROD.

The Piston is another very important part in the construction of an engine, and it conveys the power of the steam to the crank. It is composed of a piston head, on which are placed the piston rings held in position by the follower plate, and is securely attached to the piston rod. Great care should be taken in the construction of the piston rings to have them fit the cylinder perfectly tight, at the same time to have the least possible friction. Piston rings should always be made of a softer metal than the cylinder so that the greater part of the wear will be upon the rings instead of the cylinder, as the rings can easily be replaced.

There are a great many kinds of packing for piston rings, but the most commonly used at the present time are the steam packing rings. The character, accuracy in construction and condition of the piston make a great difference in the quantity of fuel consumed and the amount of power developed by the engine.

The Piston Rod connects the piston to the cross-head and is generally made of steel. Where the piston rod enters the cylinder, a steam tight joint is obtained by the use of a soft, pliable packing placed in the stuffing box, and held in position by the stuffing-box gland.

This box is kept packed just tight enough to prevent leaking, by drawing up the stuffing box gland when required. This can be repeated until the packing is all used up, when box must be repacked.

STEAM CHEST.

The Steam Chest contains the valve, and can be on either side of cylinder as may best suit the style of engine. Steam is admitted into the steam chest, and passes into the cylinder by the action of the valve.

Many engine builders cast cylinder and steam chest in one piece, while others cast them separately and bolt them together. The only advantage of the former over the latter is the absence of one less joint to keep packed.

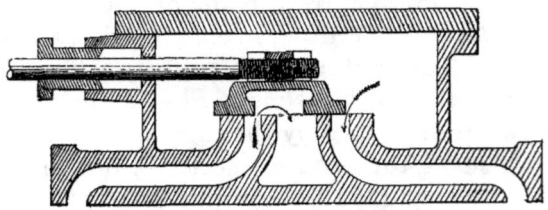
Steam Chest and Plain Slide Valve.

The steam ports are the two openings through which the steam is admitted to the cylinder.

The exhaust port is the opening through which the exhaust or waste steam passes out of the cylinder.

VALVE.

There are a great many kinds of valves used on steam engines, namely, the Corliss, Slide, Rocker, Balance, Rotary, etc., but the one most commonly used on farm engines is the plain slide valve, which has been generally adopted by all the larger engine builders in this country. It is simple in design, and when properly set does its work very efficiently. They are less complicated than others and are easily set; they are made in many different designs, but the principle of each is the same.

The slide valve is constructed to slide upon the smooth surface of the valve seat, in which are contained the two steam ports for the admission of steam to each end of the cylinder, and also the exhaust port through which the exhaust or waste steam passes out of the cylinder.

The slide valve is operated by the eccentrics, which are attached to the main crank shaft of the engine and revolve with it, the object of the eccentrics being to move the slide valve back and forth upon its seat to admit the steam alternately through the steam ports to the cylinder.

The valve gear is a most important detail and one upon which the economy of fuel in a great measure depends, and any derangement in this part of an engine causes an immediate increase in the fuel consumed and decrease in the power of the engine.

In a properly constructed valve the slide upon the seat should be reduced to the smallest possible amount, and should be so designed as to give an equal cut-off and release at both ends of the cylinder, whether working full gear or notched up.

The engine should also have the same power whether working forward or backward, and the cut-off should be as sharp as possible.

The more perfect the valve gear the more the engine can be notched up, and thus allow the steam to expand in the cylinder to its utmost. The engine which can be notched up the most is the most economical in fuel and water.

CROSS-HEAD.

The Cross-head is of cast iron and connects the piston rod to the connecting rod, and is that part of an engine where the motion is changed from vibrating to rotary. The piston rod is

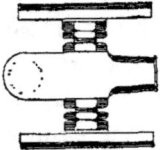
Cross-head.

fastened securely to it, while the connecting rod is attached by wrist-pin. On a V guide or bored guide engine frame the cross-head is supplied with adjustable shoe slides that can be adjusted to take up their wear and fit the guides properly, also to keep the piston rod in line. On a bar or locomotive guide engine the cross-head is adjusted by removing the liners from between the bars.

ENGINE FRAME.

The Engine Frame is the large casting which contains the bored, V shaped or locomotive guides for cross-head shoes. It also contains the pillow block for crank shaft at one end and the cylinder is bolted to the other. They are made in many different styles and shapes, but all answer the same purpose.

CONNECTING ROD.

The Connecting Rod on an engine is the connection between the cross-head and crank-pin; it is generally made of wrought iron or steel, with brass boxes at each end held in position by wrought straps. These straps are attached to the connecting rod by gibs and keys. Connecting rods are sometimes made with mortised ends to receive the brass boxes, which are held in place by wedge block and adjusting screw. The latter style is used principally on the larger makes of engines. While the connecting rod with mortised ends are considered a little the safest, the straps on the ends of rods are most commonly used. The brass boxes at ends of connecting rod are

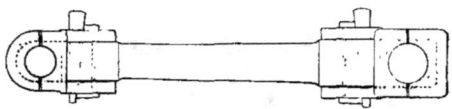
Connecting Rod.

adjustable to take up the wear by use of the gibs and keys, and they should be adjusted as frequently as there is any lost motion discovered at the crank-pin or cross-head, which will be indicated by a knocking or pounding as the crank passes over the centers.

CRANK.

The Crank is that part of an engine by which the effect of the steam acting against the piston is converted into work. There are two kinds, Side and Center crank. The term Side crank refers to a disc plate or a crank attached to one end of a shaft and in which is placed the crank pin. When the shaft extends to the right the engine is called a right hand engine, and when it extends to the left it is called a left hand engine.

The term Center crank refers to a shaft with the crank in the center, the shaft extending equally both ways and so constructed as to be very well balanced. It is optional as to which gives the best results as both kinds are used upon all sizes of engines.

CRANK-PIN.

The Crank-Pin connects the connecting rod and crank. It is made of steel, and special care should be taken to keep the crank-pin well oiled. If allowed to run dry and cut, it will soon heat and ruin both pin and boxes. If once allowed to become cut, it will be impossible to prevent it from heating.

LINK REVERSE.

The Link Reverse is composed of two eccentrics and rods, link, block and slide, also lever and quadrant for holding link in any position.

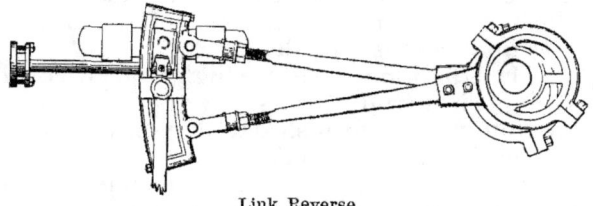

Link Reverse.

The duty of the link is to reverse the engine by simply throwing the reverse lever backward and forward. The speed of the engine can also be reduced and increased by the same operation.

RUMELY TRACTION ENGINE.

In the construction of this engine, which is of the side crank, rear gear style, it will be seen that the engine is in a different position upon the boiler from the ordinary side crank, having the cylinder forward, and the crank shaft at the rear end. The frame is of the girder pattern, with overhanging cylinder attached to one end, the pillow block bearing at the other, and is secured to the boiler by two brackets.

The engine is supplied with a Cross-head Pump, Arnold Shifting Eccentric Reverse Gear, Friction Clutch, Automatic Oiler, Governor, Large Cylindrical Water Tank on the side, and Tool Boxes upon the Platform.

The boiler is of the round bottom fire box or locomotive style, has the dome in front, and the ash pan is in the lower part of the fire-box.

It is mounted upon the traction wheels by brackets attached to the rear end of the boiler, which contain the main axle. The front end rests upon a trussed axle.

The traction wheels are high, and are of the wrought iron rim direct spoke type. The loose traction wheel is furnished with a locking device for securing it to the axle.

In the arrangement of the engine and the high traction wheels, the driving or band wheel is placed between one of the traction wheels and the boiler. The necessary fittings are furnished with both boiler and engine to keep them in good running order and perfectly safe if properly handled.

YOUNG ENGINEER'S GUIDE. 71

Rumely Traction Engine.

LINK.

The Link is that part which holds the link block and is connected at each end to the eccentric rods; it is used only on reversing engines. The link is made on a curve, so that when the link block is at either extreme end the valve is operated to its full movement. When the block is in the center of link, the valve covers both ports and prevents the ingress of steam to the cylinder.

The Link Block is attached to the slide which connects it to the valve rod. The valve rod connects the slide and valve, and where it enters the steam chest is packed in like manner to the piston rod.

REVERSE LEVER.

The Reverse Lever is that part of the valve gear connected with the link for raising and lowering it, thereby changing the travel of the valve and reversing the motion of the engine. When the reverse lever is placed in the center notch of quadrant, the lap of slide valve should cover both steam ports, preventing any steam from entering the cylinder, thus stopping the engine. In moving this lever from the center notch, it either drops or raises the link as the case may be, increasing the travel of the valve and allowing steam to enter the cylinder. When the reverse lever is thrown into the outside notch of quadrant at either end you get full travel of the valve which gives full power of engine, providing you have sufficient steam pressure.

ECCENTRICS.

The Eccentric on an engine is for the purpose of moving the valve back and forth upon the valve seat and has a throw equal to the travel of the valve. The throw of eccentric is caused by the wheel or plate being bored to one side of its true center, and generally equals one-half the travel of the valve. If more or less, the difference is caused by the use of rocker arm or similar devices for increasing or diminishing the throw of eccentric as the case may be. The eccentric is accurately fitted and fastened to the main shaft of the engine with set screws or key.

ECCENTRIC STRAP.

The Eccentric Strap is that part of the engine in which the eccentric revolves, and is attached to link by the eccentric rod. It should be kept well oiled to secure a free and easy movement to the link.

On reversing engines there are two eccentrics exactly alike, one connecting with upper end of link, the other with lower end by the eccentric straps and rods. In this case the eccentric rod that is moving the valve is the one nearest to the link block. When the lever is in the center notch the link is also in the center of its travel. In this case, both of the eccentric rods move an equal distance and the link vibrates back and forth, but as the block is in the center it gives no motion to the valve, and as the valve, having sufficient lap, covers both ports when the lever is in this position prevents the ingress of steam to the cylinder, consequently no motion.

ECCENTRIC ROD.

The Eccentric Rod connects the eccentric to the link, two being used on a reversing engine. On a simple engine only one eccentric rod is used and is connected to a rocker arm which is attached to the valve rod.

WOOLF VALVE GEAR.

The Woolf Valve Gear is used in connection with an engine to reverse its motion. It is arranged with one eccentric attached to the crank shaft the strap of which has a long arm cast on, to which the eccentric rod that moves the valve is attached. This arm is also supplied with a roller, which runs in a slot on a rocking head. This head is held in position by a box and is con-

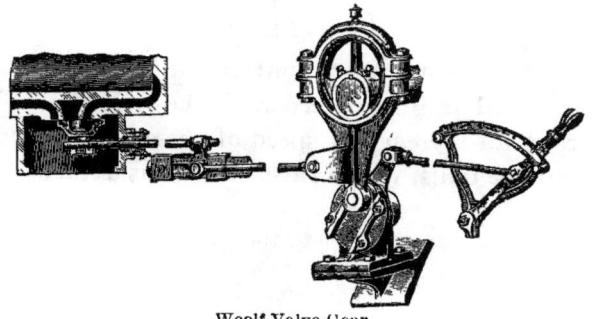

Woolf Valve Gear.

nected to the reverse lever by a rod. By throwing the reverse lever to either end of the quadrant, the position of the rocking head is so placed that the roller in the slide operates the eccentric strap, rod and valve, and the engine will run in the direction desired. By throwing the reverse lever to the opposite end of the quadrant, the position of the rocking head is so changed that it will reverse the motion of the valve, and the engine will run in the opposite direction.

The Quadrant being notched, the point of cut-off can be regulated with the reverse lever, according to the load; by placing it in the last notch in quadrant when full power of engine is required, or notching it up when doing light work, the same as with the link reverse gear.

GOVERNOR.

The Governor contains a valve so constructed and connected with the weighted balls that an increase of speed of the engine throws out the balls, which raises the arms attached to

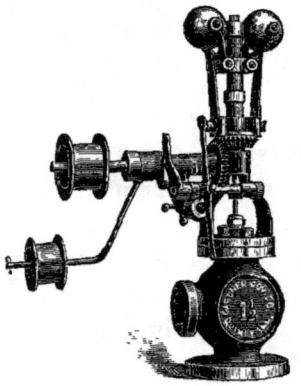

Gardner Governor.

valve rod, thereby closing the valve and thus decreases the flow of steam through the governor valve and reduces the speed of engine until the governor balls are again in their true position.

Now, when the speed decreases and the weighted balls rise above their true position, the valve opens, allows more steam to enter the cylinder and the speed of the engine increases until the engine is again running at its proper speed.

Governors are used to regulate the quantity of steam required to run an engine at a uniform speed under variation of load, and to run properly they should be kept perfectly clean and free from the accumulation of gummy substances caused by using inferior oil, which has a tendency to interfere with the free and easy movement of the different parts.

Q. If you desire to run your engine faster or slower with the throttle valve wide open, how can it be done?

A. Some makes of governors are provided with regulating screws at the top; by turning the hand nut in one direction you lengthen the valve stem and reduce the steam opening in the governor valve, which reduces the speed.

To increase the speed of the engine the handle nut is turned the opposite direction, which shortens the stem and increases the opening in governor valve, allowing more steam to enter the cylinder and the speed is proportionately increased.

The Gardner Governors are provided with a hand screw at the side for regulating the speed, as will be seen by examining the accompanying cut. It also has a Sawyer's Lever for opening the valve to its full extent, and a belt tightener.

Q. Will the handle nut stay in position after once being set to a certain speed?

A. Not unless the check nut directly over the handle nut is screwed down tight to prevent the stem from changing its position.

Q. Is a governor liable to cause trouble and fail to govern the engine properly?

A. All governors are more or less delicate in construction and must be kept clean and well oiled, the belt must not be allowed to slip, nor must it be so tight as to cause the governor to work hard. The small stuffing box that packs the valve stem should never be screwed down steam tight, as it causes too much friction on the stem and prevents the balls from operating it, and the engine will run unsteadily and spasmodically. Always allow the stuffing box to leak a little, then you know it is not too tight.

First class governors may sometimes be condemned for not regulating the engine to a uniform speed, when a good cleaning, oiling or loosening of the valve stem stuffing box nut would allow them to work perfectly.

AUTOMATIC OILER.

An Automatic Oiler is used on an engine to keep the cylinder, piston and valve lubricated, and is most essential for the safety and easy operation of these parts. It works automatically and is supplied with a glass tube through which the oil can be seen passing into the cylinder. The feed can be regulated to allow just the required amount of oil to pass into the cylinder.

Where a sight-feed automatic oiler is used on an engine there can be no excuse whatever for the engineer to allow the cylinder to run dry and cut, as he at all times can see whether oil is passing into it or not.

Q. How do you fill an automatic oiler?

A. Close valves D and E, open valve G to draw off the water. Close valve G and take out filling plug C, fill A with oil and replace plug C, then open valve D, and the flow of oil to the cylinder can be regulated with valve E.

Q. Will the oil feed as soon as the oiler is filled?

80 YOUNG ENGINEER'S GUIDE.

A. No; time must be given for sight feed glass and condensing chamber to fill with water of condensation.

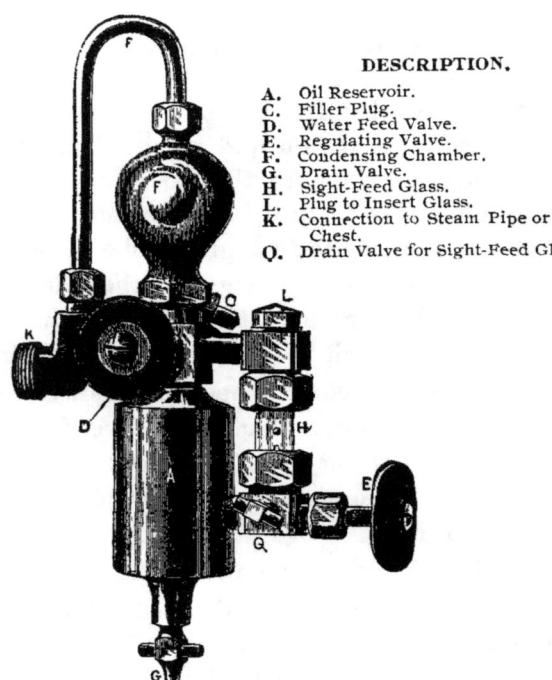

DESCRIPTION.

A. Oil Reservoir.
C. Filler Plug.
D. Water Feed Valve.
E. Regulating Valve.
F. Condensing Chamber.
G. Drain Valve.
H. Sight-Feed Glass.
L. Plug to Insert Glass.
K. Connection to Steam Pipe or Steam Chest.
O. Drain Valve for Sight-Feed Glass.

Single Connection Detroit Oiler.

Q. How is a double connection automatic oiler attached?

A. First, the steam pipe must be drilled and tapped above the throttle with ½ or ¾ inch gas tap, as may be necessary to receive the oil discharge pipe of oiler and put oiler in place. Then

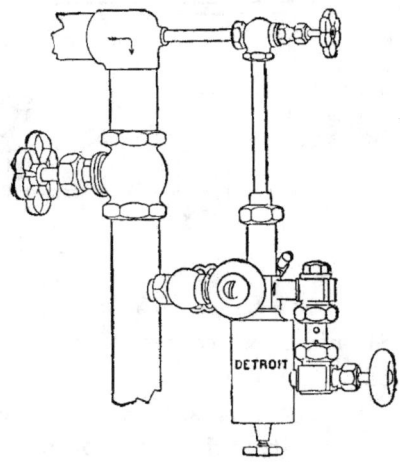

Double Connection Detroit Oiler.

tap the steam pipe about 18 inches if possible above the top of the condensing chamber and fit in a ¼ inch gas pipe for steam connecting tube, which attaches to the top of condensing chamber. (See illustration.)

Where the steam pipe cannot be tapped 18 inches above the condensing chamber, it may be tapped lower down and the steam connecting tube of required length be bent in a horizontal coil. With the single connection oiler it is only necessary to drill and tap one hole in the steam pipe.

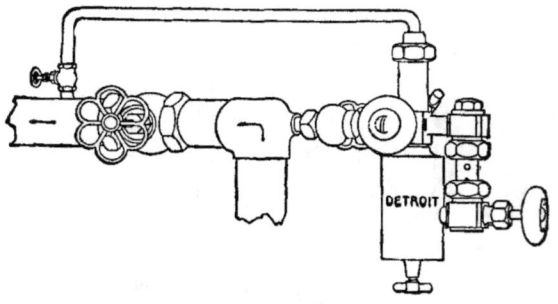

Double Connection Detroit Oiler.

Q. Should oiler become clogged, how can it be cleaned?

A. Open the valves, by which steam can be forced through it, and all the passages will be cleaned. This can be done without stopping the engine.

Q. If oiler is not in use and in danger of freezing, what should be done?

A. Leave valves D, G and E open, and all water will be drained off.

Q. If the glass tube in oiler should get broken, what should be done?

A. Shut valves D and E, remove broken glass and replace with new.

Q. How can oil be prevented from sticking to lubricator glasses?

A. A very simple remedy is to fill the glass with glycerine and let the oil feed through it.

INJECTOR.

An Injector is an automatic machine attached to a boiler, for injecting or forcing water into it and at the same time heating the water to a very high temperature, which saves fuel and prevents the danger of sudden contraction of the plates and flues. It can be used independently and is indispensable on a farm engine.

In piping an injector to boiler, use as short and as straight pipes as possible and especially avoid short turns. Take steam directly from boiler, and have a globe valve in steam pipe close to injector; have the water suction or supply pipe independent of any other connection, and it must be supplied with a globe valve close to injector. This pipe and connections must be absolutely air tight; the slightest leak will cause trouble. The discharge pipe to boiler must be supplied with a tight and reliable check valve.

If valve leaks, the injector will become hot and cause no end of annoyance. It is a good plan to put a stop valve between check and boiler in discharge pipe, so that check valve may be taken

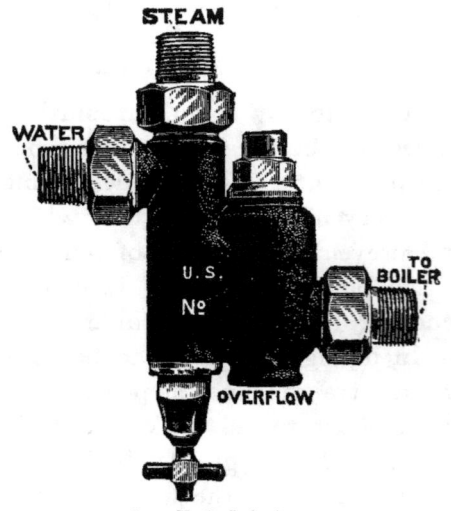

American U. S. Injector.

off and repaired, or a new one put on without loss of time. A foot of straight pipe screwed into the overflow assists in starting an injector, especially at low pressure.

Q. How do you start an injector to work?

A. To start an injector, open the suction valve wide, then open steam valve. If water appears at the overflow, close and open quickly the suction valve, opening only about ¼ of a turn if at low steam pressure, and one turn or more if at high steam pressure, regulating the water supply according to steam pressure. The injector is controlled entirely by the valve in suction pipe, or by the suction lever after the steam is turned on.

Q. Will an injector work with hot supply water?

A. An injector will not work if the water that is delivered from the tank is too hot to condense the steam.

Q. What are the principal causes of an injector not working accurately?

A. Leak in suction pipe, supply cut off by strainer being clogged, loose lining inside the the hose, leak in the stem of valve, too little steam pressure to lift, dirt in the tubes, red lead blown or drawn in through steam or supply pipe, bad check valve, not lift enough or none at all, new boiler full of grease, wet steam, obstruction in the connection to the boiler.

If injector fails, examine at all of these points before condemning. The most common trouble is a leaky suction.

In describing the method of connection and operating an injector the foregoing may have to be modified in some instances, as there are a great many different kinds and styles of injectors which operate and connect differently, but the above if followed carefully and with a little discretion will be found useful.

Bear in mind, however, that the injector does not start to work to boiler as soon as it gets the water. At first the water will run out of the overflow. At this point you start the injector working to the boiler by closing and opening the water valve as quickly as possible *with a jerk, or as nearly with one motion as you can.*

Q. How do you find the maximum and minimum capacity of injectors?

A. Injectors are controlled entirely by the suction valve after steam is turned on. To find the maximum capacity of an injector after starting, gradually open the suction valve in supply pipe until steam "breaks" and water comes out of the overflow, then start the injector again and you will know about how far the suction valve can be opened without causing the "break."

To find the minimum capacity of the injector, manipulate the suction valve in the same manner in exactly the opposite direction.

DIRECTIONS FOR OPERATING WORLD INJECTOR.

See that the injector is shut off when put on, by turning the handle as far to the right as it

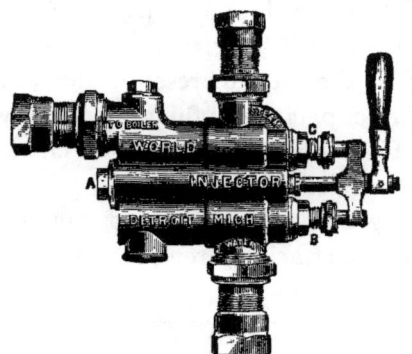

American World Injector.

will go. To start, turn handle to the left one-quarter turn; when the water appears at the overflow, turn the handle slowly to the left as far as it will go, and the injector will be working to the boiler.

If steam is high and lift long, the injector will lift the water better if the handle is turned a little less than a quarter of a turn, until the water appears at overflow—then start to boiler as before.

If you have valves in steam and suction pipes, be sure and open them before starting.

Q. How high will an injector draw its supply?

A. About twenty feet is the limit.

Q. How hot does an injector deliver water?

A. From 150 degrees to 200 degrees, according to the steam pressure and the proportions of its capacity at which injector is working.

Q. How should the jets be cleaned when they become scaled?

A. By soaking in diluted muriated acid, about one part acid to ten parts water.

THROTTLE.

The Throttle on an engine is the valve which allows the steam to enter or be shut off from the cylinder and should always be left wide open on a governor engine when running, as the governor regulates the quantity of steam required to run it at its proper speed and is much more economical.

There are different styles of valves used for throttles, such as Globe Valves, Butterfly Valves, Disc Valves, etc.

Lunkenheimer Throttle Valve.

The Lunkenheimer is a double disc valve, and is operated by the handle or rod attachment, and requires no lock or ratchet.

STEAM PUMP.

An Independent Steam Pump is virtually an engine with two cylinders, one for the steam piston, the other for the water piston or plunger, and is used in connection with a steam boiler for supplying it with water. The discharge pipe of a pump is generally connected with a feed water heater of some sort, which heats the water to a high temperature before entering the boiler, though there is a late pattern of steam pump which delivers the feed water to the boiler at about the same temperature as the injector. The cylinder of steam pump should always be

Independent Steam Pump.

well oiled before starting in the morning and stopping at night. The stuffing boxes on piston and valve rod should in all cases be kept well filled with soft and moist packing. If the packing is allowed to become dry and hard, it will cut the rod, inducing leakage and necessitating repairs. When a steam pump is not in use in cold weather all the drain, drip and pet cocks should be left open, to allow the water to run out. While most farm engines are furnished with an independent steam pump, some are equipped with what is called a cross-head pump.

A Cross-Head Pump is operated by a plunger attached to the cross-head of engine and has two valves, a supply and a discharge valve, also is supplied with an air chamber. This style of pump is available only when engine is running. Engines with a cross-head pump should always be supplied with an injector to be used in case of failure of pump to work and while engine is shut down. The cross-head pump is connected to the heater in the same manner as an independent pump.

Q. How high will the steam pump lift water?

A. A steam pump will lift or draw water about 33 feet, as with one inch area, 33 feet of water will weigh 14.7 lbs., but the pump must be in very good order to lift 20 feet and all pipes must be absolutely air tight. A pump will give better satisfaction lifting from 10 to 15 feet. No pump however good will lift hot water, for the reason that as soon as air is expelled from the barrel of the pump the vapor occupies the space, thereby destroys the vacuum and interferes with the supply of water. When necessary to pump hot water, place the pump below the supply, so that the water will flow into the valve chamber. Always have a strainer at lower end of suction or supply pipe. A pump should be set up so that it is accessible for inspection, cleaning and repairs, and so that the shortest and straightest suction and delivery pipes can be used.

MARSH STEAM PUMP.

The Marsh Steam Pump is so constructed that the exhaust steam may be turned into the suction, thereby condensing its exhaust steam and returning it with its heat to the boiler, thus heating the feed water to a high degree.

The pump is automatically regulated and can never run too fast to take suction, or should the water supply give out when the throttle valve is wide open no injury can occur to the moving parts.

The steam valve, though nicely fitted, moves freely in the central bore of the steam chest, and has no mechanical connection with other moving parts of the pump, but is actuated to admit, cut off and release the steam by live steam currents which alternate with the reciprocation of the piston.

Each end of the valve is made to fit the enlarged bore of the steam chest, and it is due to these large valve heads, which present differen-

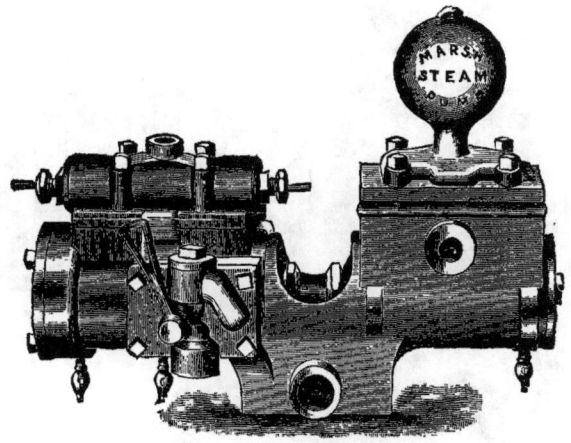

MARSH PUMP.
Capacity 10 to 35 Horse Power.

tial areas to the action of steam and the perfect freedom of the valve to move without hindrance from other mechanical arrangements or parts, that the flow of steam into the pump is automatically regulated.

94   YOUNG ENGINEER'S GUIDE.

The steam valve does not require setting, as it has no dead center and will always start when steam is admitted.

The steam piston is double and each head is

View of Marsh Steam Pump, Showing Water Valves, Steam Valves, Suction Chamber and Piston.

provided with a metal packing ring. The piston rod is made of Tobin bronze, the stuffing boxes and water piston are made of brass, and the water cylinder is brass lined. The water valves may be removed for inspection by simply taking off the air chamber.

DIRECTIONS FOR SETTING UP AND RUNNING.

Before connecting the steam pipes, blow out with steam pressure the chips and dirt in the steam pipes. Always use the union furnished with the pump. It has a gauze gasket in it to catch the dirt that may get into the valve. Before starting pump, open air cock in delivery pipe and turn exhaust lever back, away from the air chamber. Then open throttle valve wide and allow pump to exhaust into the air until it takes suction, when deflecting lever may be thrown forward toward air chamber and cold water in the pump will condense the exhaust and return it to the boiler.

If pump refuses to work, the difficulty is to be looked for in the valve chest. Do not take off the chest. The valve may be taken out and cleaned but *never* filed. *The valve must be returned through same end as taken from.* Before closing, be sure that the head is screwed tight on the valve, using the socket wrench furnished.

When the pump is stopped, pull the exhaust lever back, so the condensed steam from leak of throttle valve will not go into the pump. It is safer also in cold weather to take off head of water end. Slight but constant lubrication adds much to the regular working of the pump. Be sure there are no leaks in the suction pipe, and when water is raised more than 10 feet, a foot valve should be put in. Compress the packings on piston rod as little as possible and yet prevent the escape of steam. Before leaving the pump in cold weather, break the suction and allow it to run empty for a minute with all the cocks open, then be sure the throttle valve is closed tight. When necessary, pack the joints under the steam chest, side plate and air chamber with manilla paper or thin rubber.

HEATER.

The Heater is used on an engine in connection with a boiler for heating the feed water before it enters the boiler under steam pressure. It is usually constructed of a shell of cast or boiler iron into which live or exhaust steam is admitted. This shell usually contains a series of pipes or a coil of pipe, through which the feed water is forced by the pump, the water thereby being heated to a high temperature before entering the boiler.

EJECTOR.

An Ejector is a machine for lifting water from various depths and forcing it to various heights with steam pressure, as follows:

With 20 pounds steam pressure an ejector will lift water to a level from 16 or 17 feet below and force it to a height of 15 or 16 feet.

American Ejector.

With 60 pounds steam pressure will lift 20 feet and force to a height of 60 feet.

With 100 pounds steam pressure will lift 23 feet and force to a height of 107 feet.

An ejector may be placed in any position to suit the convenience in piping; they require but three connections, steam, suction and delivery.

JET PUMP.

A Jet Pump is a machine used for drawing water and discharging it above the surface level.

It will draw water to a level from 10 or 15 feet below and discharge it at a height of about one foot to every pound steam pressure applied.

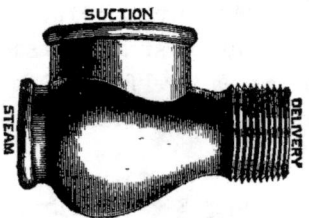

American Jet Pump.

A jet pump has three connections, steam, suction and delivery, and may be placed in any position to suit convenience in piping.

STEAM GAUGE.

A Steam Gauge is an instrument used for indicating in pounds the amount of steam pressure upon each square inch of surface of the boiler. It is very delicately constructed, and should not be tampered with after once being set to indicate correctly. If a steam gauge is found not to indicate the exact pressure in the boiler it should be sent to the factory for repairs. Never attempt to repair it unless all the appliances for so doing are at hand.

A steam gauge should always be placed on a boiler with a syphon, or by tying a knot in the pipe between it and the boiler, so that the steam may condense, thereby allowing the water to operate it. If steam is allowed to enter, the heat would tend to expand the tube in the gauge and it would indicate more than the real pressure.

Ashcroft Steam Gauge.

A steam gauge is usually constructed with a hollow, flat tube, called the Bourdon spring. This tube is bent in a simple curve and fastened at one end, the other end is free and by a simple clock work actuates the pointer which indicates upon the dial the steam pressure per square inch upon the boiler.

Q. What should be done in case the steam gauge becomes defective?

A. When the steam gauge has become broken by freezing or otherwise, and there is none on hand, the engineer may run by setting the safety valve so that it will blow off within from ten to fifteen pounds less than it is ordi-

Interior Ashcroft Steam Gauge.

narily set at, and then by careful firing, run until a new gauge can be procured, which should be done without delay.

Some engineers have been known to make a practice of running without a steam gauge.

Q. Would you recommend this method?

A. No, it must be resorted to only in case of a sudden accident and where shutting down would cause great loss. Every engineer should have an extra steam gauge on hand, to be used in case of accident to the one in use.

SAFETY VALVE.

Safety Valves, or pop valves, as they are sometimes termed, are made in many different

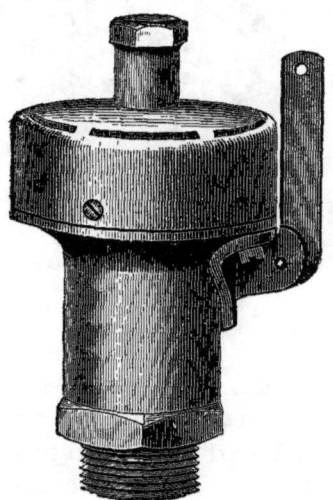
Kunkle Safety Valve.

kinds and styles, but the one most commonly used on a farm engine is constructed with a coil spring, which is adjustable, to allow the valve to pop off at a certain pressure. When the pressure exceeds this amount, it raises the valve from its seat and allows the surplus steam to escape. It should be set with the steam gauge, to allow a little margin of steam pressure over

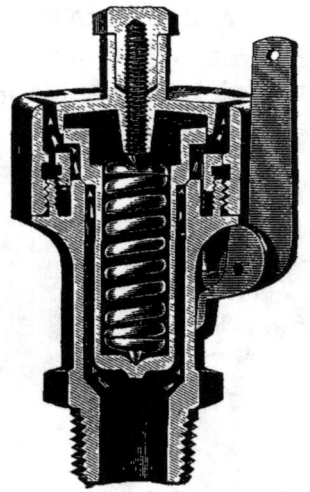

Kunkle Safety Valve, Sectional.

that which is necessary to drive the load, and should be compared with the steam gauge frequently to see that it works accurately. It is furnished with a lever for raising the valve, which should be raised occasionally to see that it operates freely.

The safety valve being set with steam gauge, the gauge should be watched when safety valve blows off. If it indicates more or less than the gauge something is wrong, the valve marked incorrectly or the steam gauge is out of order. In case the safety valve and steam gauge do not register alike, the valve should be examined to see that the valve is not stuck in its seat, and should be thoroughly cleaned of all sediment; then put back in place again and compared with the steam gauge as before. If they do not register alike, the gauge should be examined.

Q. Is not a safety valve attached to a boiler to prevent explosion and loss of life?

A. The Safety Valve is only intended to prevent an explosion from excessive steam pressure, and should never be set to hold more than the required pressure to give the rated power for which the engine is designed.

GLASS WATER GAUGE.

A Glass Water Gauge is a device attached to a boiler to show the level of water in the

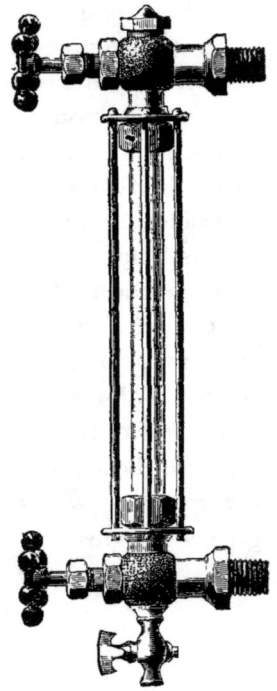

Lunkenheimer Water Gauge.

boiler. It consists of a glass tube ten or twelve inches long, with an angle valve attached at each end. The lower valve enters the boiler below the water line, the top valve enters the boiler above the water line in the steam space. The ends of the glass are made steam and water tight by means of rubber gaskets and stuffing boxes. As water will seek its own level, the height of water in the boiler will show a corresponding height in the glass, and the engineer at a glance knows just how high the water in the boiler is above the flues and crown sheet.

BLOW OFF VALVE.

The Blow Off Valve is an angle or globe valve attached to the lower part of a boiler for the purpose of blowing off the sediment accumulated by the use of impure water, and should be used more or less frequently according to the condition of the water used, but never less than once a week. There is also a surface blow off which is attached to the boiler at about the safe water level for the purpose of blowing off the scum which accumulates on top of the water. This scum should be blown off once or twice a day.

GAUGE COCK.

A Gauge Cock is a stop cock attached to a boiler to ascertain the height of water in the boiler. There are generally two and sometimes three gauge cocks attached to a boiler. The lower one enters the boiler as low down as it is deemed safe to allow the water to get, while the upper one enters the boiler sufficiently high to

Lunkenheimer Gauge Cock.

avoid getting too much water. Where three gauges are used, the middle one enters the boiler at about the proper water level. They should be opened frequently to keep them free from corrosion, being sure to close them tightly to prevent leaking.

CYLINDER COCKS.

Cylinder Cocks are used in connection with the cylinder to allow the water accumulated by the condensing of steam in the cylinder to escape, and should be opened every time the engine is started or stopped. This should never be neglected, as great damage may be caused by the

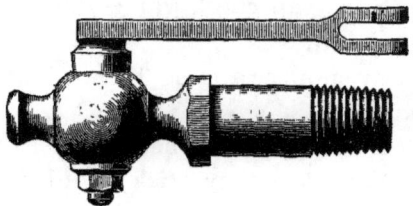

Lunkenheimer Cylinder Cock.

breaking of the follower plate or cylinder head. They should also be left open when engine is shut down, and at night in cold weather to prevent freezing and the consequent damage.

CHECK VALVE.

A Check Valve is the valve on the feed pipe between pump or injector and the boiler. The check valve is so constructed that the pressure of the feed water from the pump or injector lifts the valve from its seat and the water passes into the boiler. Immediately after the pressure from the pump or injector is released, the pressure from the boiler closes this valve and prevents the water from being forced back into the pump or injector. There are two check valves used in connection with a pump, one on suction, the other on delivery pipe.

Q. How can check valves that get stuck open be closed?

A. By simply tapping them slightly with a light hammer.

Q. Should this be practiced?

A. No, when they stick at all they should be opened and thoroughly cleaned as soon as pressure can be shut off.

COMPRESSION GREASE CUP.

Compression grease cups are used extensively on engines for lubricating the crank pin and wrist pin, also pillow block bearings for crank shaft; in fact this method of lubrication for all kinds of bearings and journals is becoming universally adopted.

In construction the Besly Bonanza Cup is simple and durable. The outer casing is threaded and fits tightly over the bottom part of cup. After the cup has been put in place, remove the top part, fill it with Helmet oil or grease and screw it on the lower part for two or three threads. The cup is then in working order.

To insure a plentiful supply of grease to the journal or bearing, all that is necessary is to give the top part of cup an occasional turn.

C. H. Besly & Co.'s Helmet Solid Oil is for use in compression cups, and is a perfect lubricant, in fact it is said to be the best.

MINNEAPOLIS TRACTION ENGINE.

The engine is supplied with a return-flue, straw, wood and coal burning boiler. The shell of the main flue is cylindrical in shape, but tapers toward the front end. By this arrangement the return flues can be set lower at the front end, which it is claimed has some advantage in protecting front end of flues when ascending steep hills.

The steam used is superheated by being conducted through a pipe which extends from the top of the dome on the inside of boiler through the front head and smoke stack to the steam chest.

The engine is of the Side Crank, Side Gear style, the frame of which is the girder pattern with bored guides, and has an overhanging cylinder attached to one end, and contains the pillow block bearing at the other.

It is fitted with the Woolf Valve Gear for reversing, Friction Clutch, Cross-head Pump, Injector, Syphon for filling water tank on front end, and a large Foot Board with Tool Boxes attached. The wheels are of steel, and the traction or drive wheels are furnished with malleable mud cleats extending diagonally across the entire width of tire.

The boiler is safe with proper care, economical in fuel, and the engine moves over the road rapidly. Although simple in its general construction, it has all the appliances and fittings necessary on a traction engine.

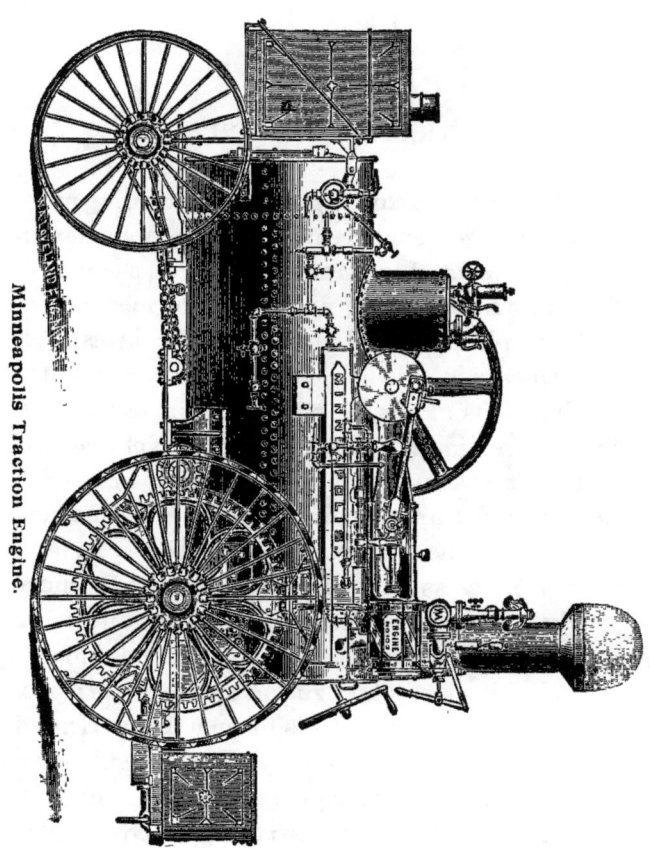

Minneapolis Traction Engine.

TRACTION ENGINES.

Traction Farm Engines are becoming more generally used each year, and to supply the increasing demand for traction engines, the manufacturers have spared no expense or mechanical skill to place upon the market the latest improvements in this line, and a great many kinds of this class of engine now made in the United States are simply perfect. They travel over the roughest roads, up and down steep hills with heavy loads, and the engine is at all times entirely under the control of the engineer. In fact, there is no class of engines that has had a more marked advancement toward perfection in the past few years than the farm traction engine.

As this class of engines in the majority of cases goes into the hands of men inexperienced in the handling of machinery, they are subjected to the very hardest usage and neglect, which, of course, hastens their destruction. Every purchaser of an engine should acquire sufficient knowledge of the operating and handling of it so that he will know when it is properly cared for.

No engine has to run at more variable speeds than a traction engine. It is very important for this reason that the steam ports should be of sufficient area to admit of a very high piston speed, and allow the steam to follow the piston at the necessary velocity. Small ports are useless, as when the link is notched up, and the travel of the valve thereby reduced, the openings are too cramped for the steam to pass in and out of the cylinder comfortably. The result is, that the slide valve is forced off its seat and the engine primes as soon as any great speed is attained. It is easy to tell by the sound of the exhaust if the ports are rightly proportioned, and whether running at high or low speed, the engine should give a clear and distinct exhaust at every stroke of the piston.

GEARING.

The power of the traction engine is transmitted to the traction wheels by a series of gearing, all of which should be made from accurately cut patterns to insure the teeth meshing perfectly to prevent them from cutting and grinding out, and of sufficient strength to withstand the very rough usage to which they are subjected. The gearing of an engine should never be run without first greasing them thoroughly with a good quality of solid oil or axle grease, for if once you allow the teeth to cut, it will be impossible to stop them from cutting, the result being they will soon be ruined. The gearing on traction engines are placed in many different positions for transmitting the power to the traction wheels, as will be seen by examining the different illustrations given in this book.

On a side gear engine the power is transmitted from a small pinion on the main shaft to an intermediate gear, from this gear to the differential gear on cross shaft which is placed under the cylindrical part of the boiler against the fire box. To this shaft are keyed two small pinions at each end, which mesh in large spur gears fastened securely to the traction wheels.

On a rear gear engine the power is transmitted from the small pinion on crank shaft to large spur gear attached to one end of cross shaft, which crosses the boiler at the rear end. To the other end of this shaft is attached a small pinion that meshes in the large differential gear which is attached to traction wheel and main axle.

DIFFERENTIAL GEAR.

The gearing on a traction engine must be so designed as to allow one of the ground or traction wheels to run faster than the other, when turning engine either to right or left on the road. To accomplish this, the Differential Gear is made use of, and answers all purposes admirably.

Differential Gear.

The Differential Gear on many styles of engines is attached to the cross shaft, while on others it is secured to the main axle, and, as generally constructed, consists of one large spur gear, having three or four bevel pinions placed in it at equal distances apart and from the center of main gear. These pinions revolve loosely on pins secured to the gear, and the spur gear runs loosely on the cross shaft or axle. At each side of the spur gear are placed bevel gears, meshing into the bevel pinions, one of which is keyed fast to the shaft or axle, while the other is firmly bolted to small pinion that drives the traction wheel, or to hub of the traction wheel, which also runs loosely on its axle.

This device allows one drive wheel to remain idle while the opposite wheel may revolve as fast as is required to make the turn.

Differential Gears should be kept well greased with solid oil or axle grease to prevent the cogs from being cut and wearing away rapidly.

FRICTION CLUTCH.

This attachment on a traction engine is almost indispensable as it allows the engineer to give the whole power of the engine instantly to the traction gear in getting the engine out of bad places, or to move the engine backward or forward with so little apparent effort as to be almost imperceptible, while the engine may be running at full speed; also for tightening the main drive belt when attached to machinery without stopping the engine.

The engine fly wheel has a heavy rim, one half of which is turned or faced accurately on the inside to form a bearing for the wooden friction shoes. These wooden shoes are held in

Friction Clutch.

an iron frame one end of which is loosely bolted to driving arm while the other end is connected to the toggle levers, so that when force is applied to engage the friction to start the engine the wooden shoes are brought into contact with the rim of the fly wheel.

The fly wheel is keyed to the crank shaft

while the driving arm with driving pinion attached runs loosely upon the shaft.

Upon the hub of the driving arm is placed a sliding sleeve to which the toggle levers are attached, the whole being operated by a lever from the platform near the reverse lever.

The adjustment to take up the wear of the wooden shoes is accomplished by means of the double nuts which form a part of the toggle levers.

Many other styles of clutches are made by engine builders adapted to their particular style of engines, but they are all upon the same principle and for the same purpose, although some have three in place of two friction shoes operated by the same lever.

When operating a friction clutch, always draw or push the lever over gradually. By doing this the engine will start slowly and easily, while if the lever is jammed over suddenly, the engine will start with a jerk, which is liable to damage the traction gear. The latter should never be done unless absolutely necessary to get the engine out of bad places on the road.

QUESTIONS WITH ANSWERS,

Concerning the Operation and Care of Steam Engines and Boilers.

Q. What should be done first, after receiving a new engine, to prepare it for running properly?

A. If a traction or farm engine, remove the box containing the fittings and tools, from the fire box, and see that the grates are in their proper places. Also take out the tools which are packed in the smoke box at front end of boiler. Then, with waste or rags well saturated with kerosene, turpentine or benzine, wipe off all the grease that the manufacturer has put on to protect the bright work from rusting. After this has been thoroughly done, clean every oil hole and bearing found upon the engine, of all dirt and cinders. Special attention should be given to this, as if dirt and cinders are allowed to remain, the bearings will cut and heat.

Q. After this is done thoroughly, what next?

A. Take all the fittings from the box and clean them carefully, fit each oil cup to its proper place and screw them in tightly with a wrench, to prevent them from working loose and falling off while engine is running on the road.

Fill all cups with good oil, lard oil for bearings, good cylinder oil for the automatic oiler, which oils the cylinder and valve, and solid oil for both grease cups at crank and cross head. Then put the steam gauge, the glass water gauge, gauge cocks, safety valve, whistle, surface blow-off and blow-off valves, cylinder cocks, governor belt, etc., in their proper places; all fittings should be screwed up tight with a wrench. Examine the stuffing boxes and see that they are all well packed and cleaned.

Q. The fittings all being attached, what next?

A. Proceed to fill the boiler with water by unscrewing the cap from filling plug located on top of boiler near the steam dome, screw funnel on plug, and fill boiler with as clean soft water as is obtainable.

Q. How much water is required in the boiler before starting fire?

A. Fill the boiler until the water shows about one and one-half inches in the glass water gauge, or have a free flow of water from the lowest gauge cock.

Q. After the boiler is filled with water to the proper level, what next?

A. Start a moderate fire with dry wood in the furnace or fire box, and open the draught damper wide. Add fuel slowly, and while steam is being raised take your oil can and wrench and examine the engine thoroughly at all its parts. See that every screw and bolt is tight and that none of the oil holes have been overlooked.

If a traction engine, examine all the gearing and see that all gears, axles and bearings are thoroughly greased and oiled—grease for gears and axles, oil for bearings.

If firing with coal, keep the grates well covered with a thin layer. Do not throw in large lumps or too much fresh coal at one time. A thin fire lightly and frequently renewed, is the most economical.

Q. Is the natural draught of the boiler enough to enable steam to be raised quickly?

A. No. The boiler and water being cold the fire will not burn briskly, but as soon as steam pressure shows upon the steam gauge, turn on the blower, which will force and increase the draught; then with good fuel, any desired steam pressure can be raised quickly.

Q. Must the blower be used when the engine is running to keep up sufficient steam pressure?

A. No. When the engine is started, the exhaust steam is discharged from the cylinder first through the heater, then into smoke-stack, producing the same effect as the blower.

Q. If the boiler steams too fast, what should be done?

A. Simply close the damper. Do not open the fire door, as the fire door should never be opened unless absolutely necessary, nor should it be kept open longer than is needed, as the cold air admitted through it injures the boiler and is wasteful of fuel.

Q. After sufficient steam is raised, how do you proceed to start the engine?

A. Before turning steam on the engine, go to the fly wheel and turn it a few times to see that everything is all right and no obstacle in the way to prevent the engine from running when steam is applied, being sure to leave the crank pin off the center to enable the steam to start the engine when throttle valve is opened. Next, open both the cylinder cocks, then the throttle valve just a trifle to allow a little steam to enter the cylinder, to warm it and expel the water of condensation. Then open the throttle gradually, and if everything is right, the engine will move off faster and faster until the proper speed is attained. After engine is thoroughly heated and is working dry steam, close the cylinder cocks and set the automatic oiler to work.

Q. How do you obtain the proper speed, and how is the engine made to run steadily with the steam pressure so varied?

A. The proper speed and steadiness in running is maintained by the use of the governor, which receives its motion from the engine shaft by means of a belt.

Q. Are the bearings of a new engine liable to heat when first started up?

A. Not if proper attention is given to them. When starting a new engine the first time, it should be stopped frequently and the moving parts and bearings carefully examined. Feel of all the bearings, the link block, the eccentrics, crank pin, cross-head, etc., to ascertain if they are heating. If they are, slacken up the boxes a little, but not enough to make them knock or pound. Always be careful not to loosen or tighten bearings or keys too much; just a trifle at a time, but do it often, until the bearings and boxes run cool, but tight. If this is done carefully, the engine will run smoothly and quietly.

Q. After the engine is started, what should be done next?

A. Fill the tank on the engine with water and start the injector to work, so that the proper level of water may be kept in the boiler. The independent pump if used, should now be fitted, connected and tried, to see if it is in proper shape to feed the boiler. If a cross-head pump is used, it should be fitted and attached to water supply with the suction hose. In this case when the engine is running, the pump can be regulated to supply the required amount of feed water.

Q. How is the boiler supplied with water while the engine is stopped?

A. By the independent pump or injector.

Q. Has the independent pump sufficient capacity to supply the boiler with water under all conditions?

A. Yes, always, when running at a reasonable speed.

Q. Why should an injector be furnished if the pump will supply the boiler?

A. Many times through carelessness or otherwise the pump is prevented from working by dirt, straw, chips and other obstructions which find their way into the pump and hold the valves from their seats. In this case it is necessary to take the pump apart and remove the obstructions wherever found, which would necessitate stopping the engine and allowing the steam to go down, involving a large loss of time. Whereas if the engine is also supplied with an injector, should the pump fail, the injector can immediately be started and the pump examined at leisure without loss of time and avoiding all danger of explosion.

Q. Should the supply of feed water be continuous while the engine is running?

A. Yes. Gauge the speed of the independent pump so that it will furnish the required amount of water to the boiler. Regulate the feed of a cross-head pump, by the suction valve. By so doing, the boiler steams easier, the flues are not so liable to leak, and a uniform steam pressure can be easily maintained more economically.

Q. How is a boiler supplied when engine is in motion?

A. By the independent or cross-head pump.

Q. When should the injector be used in preference to the independent pump?

A. There being no exhaust steam when engine is not running, no benefit is derived from the heater. Now, as cold water should never be forced into a hot boiler the injector becomes of great value, as it heats the feed water to a very high temperature before it enters the boiler.

Q. Is there any independent steam pump made that heats the feed water before it goes into the boiler?

A. See Marsh Pump description.

Q. When engine, pump, and injector are found to be working properly, what next?

A. If a traction engine, the engine should be reversed several times. This can be done by throwing the reverse lever forward and backward, to ascertain whether the valve is so set that engine will run equally well both ways; then the traction gear may be tried. If engine is supplied with a friction clutch, by simply pressing the clutch lever gradually until the friction shoes take hold, the engine will start slowly upon the road. This can be done while engine proper is running at full speed.

The clutch lever should be held in one hand when first starting, so that in case of anything being wrong with gearing, it can be stopped immediately by quickly loosening the lever. With the other hand, the steering wheel should be operated to guide the engine upon the road. When all is found to be working properly, and you wish to run the engine any distance, the clutch lever should be placed in notch provided for it; this will hold the friction shoes securely to the wheel, and the engine will move along the road at full speed.

Q. If the engine has no friction clutch, how do you proceed to start the gearing?

A. Stop the engine and place the reversing lever in center notch, slide the spur pinion on main shaft into gear and open the throttle valve wide; then with the reverse lever in one hand (the steering wheel in the other) you can start engine upon the road by throwing the lever backward or forward, which should be done gradually at first, so that engine will start slowly. If all is right, by throwing the reverse lever in the last notch in quadrant, the engine will travel its full speed upon the road.

Q. How should a traction engine be first started upon the road, forward or backward?

A. Always forward, as you can see where you are going and can guide the engine more easily.

WATERTOWN HIGH SPEED ENGINE.

The frame of this engine is very heavy, with longitudinal and cross ribs securely bracing it. It forms the lower guide for cross-head, and contains the pillow block bearings at the front end. It also forms the front cylinder head, to which the cylinder is bolted. The working parts are placed as low in the frame as is possible, so that the strain is brought in line with the line of greatest resistance.

The double disc center crank shaft allows of two small heavy band wheels, one of which has the shaft governor attached to the inside, that operates the valve automatically to give the point of cut-off in accordance with the variation of load.

The valve is of a special design, and so constructed as to admit steam to the cylinder port through four different openings. It also exhausts steam through four different openings at once.

The engine frame rests its whole length and is securely bolted to the sub-base, which is bolted to the floor, and the smaller sizes need no elaborate foundations.

This style of engine ranges in size from 35 to 350 horse power approximately, the speed of the smaller ones ranging from 275 to 325 revolutions, and of the larger from 160 to 185 revolutions, and are used extensively where high speed is required, and where the load is of an intermittent character.

Watertown High Speed Engine.

Q. How should you guide a traction engine?

A. There is no fixed rule for guiding a traction engine upon the road. It must be learned by experience. Good judgment is required to make a success of it. One man should always handle both reverse lever and steering wheel when guiding an engine.

Q. How should the steering chains be put on a traction engine?

A. The chains should be so put on, that when the steering wheel is turned to the right, the engine turns to the right; when wheel is turned to the left, the engine turns to the left.

REVERSING AN ENGINE.

To reverse the motion of a plain Slide Valve engine, remove the cover of steam chest and place the engine on the dead center. Observe the amount of lead or opening that the valve has on the steam end, then loosen the eccentric and turn it around on the shaft, in the direction the engine is wanted to run, until the valve has exactly the same amount of lead at the other end; then turn the engine to the opposite center, to determine whether the lead at this end is the same as at the other; then place the crank at half stroke top and bottom, and see that the port openings are equal in both positions, and replace cover.

Q. What is dead center?

A. The dead center of an engine is the point where crank and piston rod are in an exact line.

Q. What is a half stroke?

A. It is the point reached by the piston after traveling exactly one-half its travel.

Q. What is the meaning of "lost motion"?

A. Lost motion is looseness of the connecting rod at crank or wrist-pin, or looseness of the link or link block, cross-head and guides, main crank shaft and pillow block, caused by friction and wearing away of the metals and the neglect to take up the wear.

The lost motion in all parts of an engine should be taken up frequently, to keep it running smoothly, but care must be taken not to tighten bearings or keys too tight.

Q. What is lap and lead?

A. Lap is the position of the valve which extends or laps over the edge of the ports when the valve is in its central position. That on the inside of the D is the inside or exhaust lap, while that at the ends of the valve is the outside lap and affects the admission and cut-off.

Lead is the amount of opening which is given to the port by the valve when the engine is on the center.

Lead on a valve is the admission of steam into the cylinder before the piston completes its stroke.

Q. How much "lead" should a valve have?

A. There is no general rule for the amount of lead that would be best suited for all makes of engines. It must be determined by the design or construction, speed and work required, to produce the best results for economy and quietness in running.

Q. What is a throttle engine?

A. A throttle engine is one in which the speed is controlled by throttling the steam with a governor, as opposed to an automatic engine in which the speed is regulated by varying the point of cut-off at the valve.

Q. What is the difference between a stroke and a revolution?

A. A stroke is the movement of the piston, from one end to the other of cylinder. A revolution takes two strokes of piston.

Q. How are steam packing rings put on the piston head?

A. Remove the back cylinder head and take the piston and rod out of the cylinder, and stand it "head up." Then place the inside of the ring, opposite to the opening, against the side of the piston head next to you; gradually press the ring open with your hands, and it can be easily slipped over the head and put in place.

Q. How is a piston put into cylinder?

A. It is always entered from the back end in horizontal engines. The rings (if steam packing) should be carefully placed in position and compressed by the hand if piston is small, but if large, curved blocks of wood or a band of sheet iron can be used to support them until they enter cylinder. When the piston rod passes through the stuffing box, it should be supported at outer end to prevent cutting.

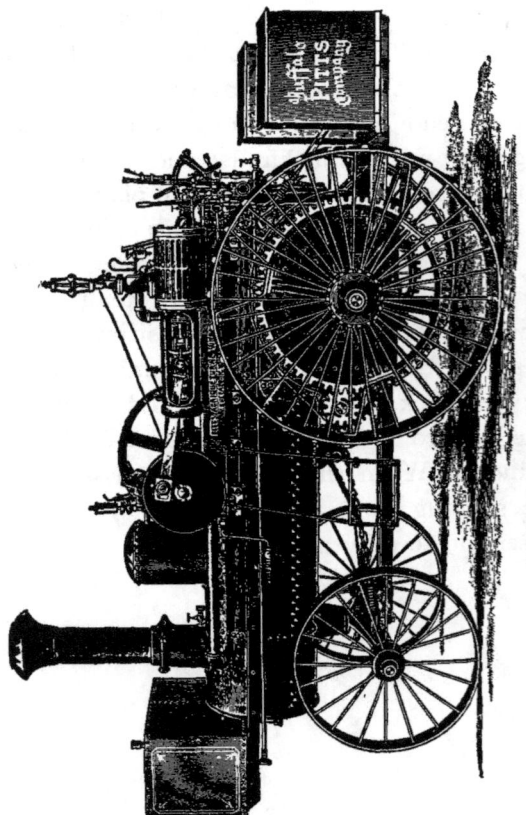

Buffalo Pitts Traction Engine.

BUFFALO PITTS TRACTION ENGINE.
COAL AND WOOD BURNER.

This engine is of the Side Crank, Side Gear construction; the larger sizes of which are of the Center Crank, Side Gear type.

The engine frame, shown in cut, is cylindrical in form with bored guides and large lateral opening; it also forms front head for cylinder.

The long heater attached to side of boiler, forms the engine bed to which the engine frame cylinder and pillow block bearing for crank shaft are securely bolted.

The piston connection with crank disk is a solid connecting rod without straps or keys.

Steam is admitted to the engine by a quick opening throttle valve.

This engine is fitted with a Pickering Quick-acting Governor, Woolf Reverse Valve Gear, Swift Sight Feed Lubricator, Three Point bearing Friction Clutch, Moore Pump, Penberthy Injector and a 100 gallon water tank on front of boiler.

The boiler is of the round bottom locomotive fire box type with high dome at front end. It is mounted upon the traction wheels with axle arms attached to brackets bolted to side of fire box. The straight stack has a spark arrester on the inside.

The traction wheels are of the steel rim and spoke design with malleable iron grouters riveted diagonally across the rim.

The band wheel, steering wheel and friction clutch lever are on the right side of boiler and the platform is on springs. In general the engine has every modern appliance for strength, convenience and economy in fuel and labor. It is easily handled, safe, with proper care, and will furnish abundance of power.

BUFFALO PITTS TRACTION ENGINE.
RETURN FLUE STRAW BURNER.

The general design and construction, given in the foregoing description of the Buffalo Pitts coal burning traction engine, will apply to the straw burning engine in every particular with exception of the boiler which is of the return flue round bottom fire box type with high dome at front end. The boiler is lagged and jacketed and the stack, which is located at rear of boiler, has a screen spark arrester at the top.

This engine is intended to burn straw, but will burn either coal or wood with equal efficiency and is said to be very economical in fuel.

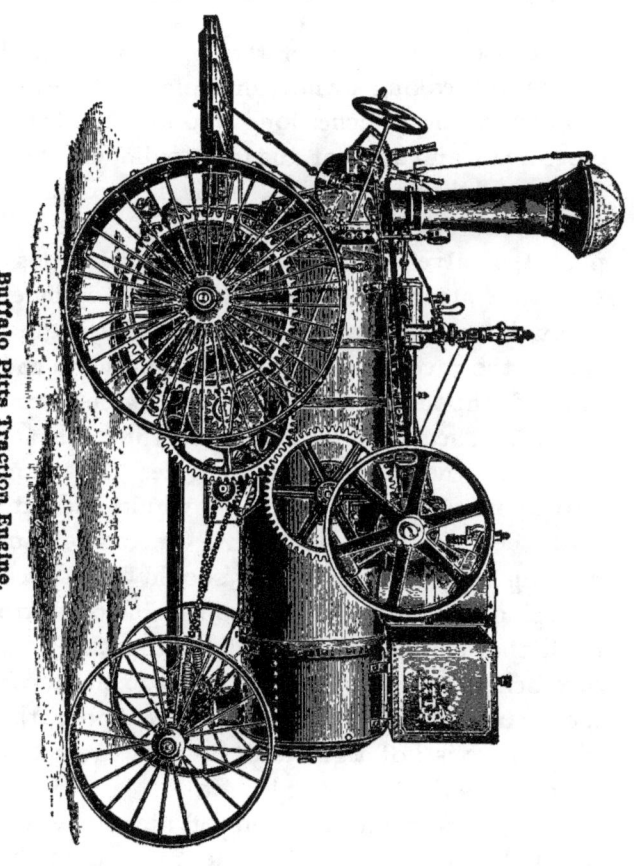

Buffalo Pitts Traction Engine.

TESTING PISTON RINGS AND VALVES.

To ascertain whether the piston rings and valves are leaking or not, first place a block of wood about four inches long upon the guide for cross-head and against the front cylinder head. Then turn the engine "over" until the cross-head comes back tight against the block. This will place the valve of a simple engine in such a position that steam would be admitted to cylinder through front steam port. If a reversing engine, throw the reverse lever in the last notch in end of quadrant that would allow the engine to run "over". Then take off the back cylinder head, and open the throttle valve wide. If steam blows out past the piston, it would indicate that the rings were not tight. In case steam packing rings are used, they should be re-fitted or replaced by new ones. If adjustable rings are used, they should be adjusted to stop the leak, care being taken not to get them too tight. If the steam blows out through the back port, the slide valve is not tight and it will require re-fitting, as will also the valve seat.

If a very small amount of steam blows through, new rings or adjustment would not be necessary, as the waste of steam or back pressure resulting from it would amount to very little.

This test should be tried occasionally with a full head of steam on, to assure of the piston rings and valve being tight, as leaky pistons and valves are very wasteful of steam, sometimes causing priming, and greatly diminishing the power of the engine. Never run your engine with leaky piston or valve. Have them properly fitted at once.

After making the test, and repairing if necessary, replace the cylinder head, and be sure to remove the wood block from the guides.

KNOCKS OR POUNDS.

The Knocks or Pounds of steam engines are frequently caused by lost motion in the crank and wrist pin boxes, valve rod and valve, crosshead and link; looseness of the piston rod, pillow block or main bearings, follower plate, or eccentrics; the slide valve having not enough or too much lead, the exhaust being cut off too soon or too late, shoulders being worn in each end of the cylinder by the packing rings not traveling over the counter bore at the end of each stroke; or shoulders may be worn in the guides by the cross-head slides, or they may not be adjusted properly to fit the guides; boiler may foam, causing the water to be drawn over in cylinder; the piston rings may leak, thus causing cushioning; and as the crank approaches the centers, steam occupies the space between the cylinder head and piston, causing a tremendous strain upon the engine; the piston rod being packed too tight, the boxes and pins being worn flat or oval; the key in driving pulley may be loose.

TO REMEDY KNOCKS OR POUNDS OF A STEAM ENGINE.

While it is hardly possible to prescribe a remedy for all cases, if the following practical methods are closely followed they will be found to be very useful, although in many instances the remedy must be determined by the circumstances of the individual case.

The knock or pound of boxes in connecting rod at cross-head or crank pin or the valve rod, may be remedied by taking out the boxes, and filing off the top and bottom inside edges sufficiently to allow them to just come together and not fit the pin too tightly. In replacing them, be careful not to key them up too tight.

Where there is not sufficient draught in the key or gib, place a liner in front or behind the boxes.

Knocks or pounds in the link may be remedied (if the link block has too much play in the link) by reducing the liners in each end of the link enough to fit the block properly.

The knocks or pounds of piston are caused by the rod becoming loose in the head; and if allowed to continue, will destroy the fit of the rod in the hole. It may be remedied under such circumstances by removing the rod, re-boring the hole and bushing it perfectly true, and refitting the rod.

Knocks or pounds of follower plate are generally caused by dirt accumulating in the hole, which will not allow the bolts to enter far enough to take up the lost motion of the plate, or the bolts may be too long. To remedy this, remove the accumulation of dirt from the hole, or shorten the bolts.

The knocks or pounds in main crank shaft, if caused by the bearings being worn oval or out of round, may be remedied by removing the shaft; true it up in a lathe, and refit or re-babbitt the boxes.

The knock or pound in eccentrics is generally caused by the eccentric straps being too loose upon the eccentrics, which can be remedied by reducing the liners in the straps to allow them to fit perfectly.

The knock or pound in slide valve caused by being improperly set, may be remedied by taking off the steam chest bonnet and re-adjusting the valve so as to give the same amount of lead at each end of stroke. This being done, and the valve well proportioned and the connections properly fitted, there should be no knocks or pounds from this cause.

The knock or pound in cylinder caused by shoulders being worn in it, can be remedied by re-boring the cylinder, being sure to make the counter bore of sufficient depth to allow the piston rings to overlap them at the end of each stroke.

The knock or pound on guides caused by shoulders being worn on them, is remedied by planing the guides and making the shoe slides of sufficient length to overlap the guides at either end when crank is on the center.

The knock or pound caused by the cross-head slides not fitting the guides properly, may be remedied by adjusting them both top and bottom to fit the guides closely, being careful not to get them too tight, which causes undue wear and strain upon the frame.

The knock or pound caused by wrist pin or crank pin becoming worn flat or oval, may be remedied by filing them perfectly round.

The knock or pound caused by the piston leaking, which causes cushioning, can be remedied only by having a tight piston.

The knock or pound caused by the driving pulley key being loose, can be remedied by driving the key in its seat; or if a defective key, replace by a new one perfectly fitted.

If the knocks or pounds are caused by lost motion in any of the revolving, reciprocating or vibrating parts of an engine, they may be detected and located by placing the finger upon the different parts while the engine is running very slowly or worked back and forth by hand.

HEATING OF JOURNALS.

The heating of journals and reciprocating parts of an engine may be attributed to the following causes:

Improper proportions and fitting, unsuitable

material, want of homogeneity between the metals of which the journals and bearings are composed, the revolving or reciprocating parts not being in line, the boxes being keyed up too tight, sand or grit getting into the journals, improper lubricating, etc. The last mentioned cause is very complicated, as the conditions of weight of load, area of surface subject to pressure, velocity of movement, etc., must be taken into consideration.

To remedy the heating of journals which is caused by the revolving or reciprocating parts not being in line, the engine or shaft must be put in line.

When caused by the boxes or bearings being too tight, they must be loosened a very little at a time until bearings run cool. Apply plenty of good oil.

Clean the boxes and journals thoroughly, and see that the oil-holes are not stopped up; also see that the oil-cups are clean, to assure of the oil getting to the bearings freely.

Oiling frequently, using a little oil at a time, gives the best results and is the most economical.

PACKING PISTON AND VALVE RODS.

When the piston or valve rod of an engine or pump needs re-packing, take off the stuffing box gland, remove all the old packing carefully, and replace with new.

If a patent packing is used, it should be cut in suitable lengths diagonally across the packing, making the angle of one end opposite from that of the other, so that when ends are brought together they will make a splice joint. The joint of each ring of this packing should be placed at opposite sides of the rod, and the stuffing box filled, the gland replaced and screwed up just tight enough to stop leakage. If hemp packing is used, take about the amount required and pick it to pieces, removing all sticks, lumps or hard substances. Then twist it into three compact cords, saturate well with oil or tallow, and braid the cords together tightly. Then wind this braid around the rod until stuffing box is full, replace the stuffing box gland, and screw up as before described.

Care should be taken not to screw the packing in stuffing boxes too tight, as it not only increases the friction on the rod and diminishes the power of the engine, but will have a tendenc to flute the rod. If the rod is once fluted, it will be very difficult to stop leakage at this joint.

When stuffing box of water piston of pumps needs re-packing, the same rule will apply, with the exception that little or no grease or tallow should be used upon the packing.

Always keep piston and valve rod packing in a clean place, as any dirt or gritty substance that may become attached to it will have a tendency to cut the rod.

SETTING A PLAIN SLIDE VALVE.

First, take up all "lost motion" and place the engine on the center. This is done by putting the wrist-pin, crank-pin and center of the main shaft in line. To do this accurately, turn the engine until the cross-head is about half an inch from the end of its stroke, and mark the position of the cross-head on the guide.

Place a marker against the edge of the fly-wheel, and make a mark on the fly-wheel opposite the marker, then turn the engine until the cross-head completes the stroke and comes back to the mark made on the guide.

The crank will now be as much below the center as it was above before. In this position make another mark upon the fly-wheel opposite the marker.

Now, midway between the two marks on the fly-wheel, when turned opposite the marker will put the engine on dead center. Next, remove steam chest cover and place the eccentrics about one quarter turn ahead of the crank in the direction the engine is to run. If the engine is to run "over", place the throw of eccentric up. If it is to run "under", place throw of eccentric down.

Then set eccentric carefully at such a point that the valve will have just commenced to open say $\frac{1}{32}$ of an inch on the end that should be taking steam. If there is a rocker arm used which reverses the direction of the motion, i. e., making the valve stem move in the opposite direction from the eccentric rod, the eccentric must be set behind the crank, when the engine is to run "over", in order that the port may open as the engine turns forward; but if a rocker arm is used merely to multiply the motion without changing the direction, proceed as though there was no rocker arm at all. Next, measure the "lead" which you have given to the valve at end which you have set. This is easily done by pushing a wedge-shaped stick or piece of soft wood into the port opening. The edge of the valve and port will mark the distance it goes in. Turn the engine upon the other center, which will be found as before described, and see if the lead is the same at both ends of the valve. If it is, the engine is properly set. If it is not, move the valve on the stem towards the end having the greatest amount of lead, a distance equal to one-half the difference in the leads. If the equalized lead is more than is necessary, set the eccentric back a little.

There are numerous methods of attaching the stem to the valve. A common way is with jam-nuts. With this arrangement it is only necessary to turn back the nuts on end towards which the valve is to be moved, an amount which will allow the given movement; then, turn the other nuts until the valve is forced into place to travel equal distance both ways from its center position. When the stem screws directly into the valve, the connection to the rocker arm or guide must be taken apart, and the stem screwed into or out of the valve enough to give it the required position. After the valve is set, replace the steam chest cover, and secure the eccentric perfectly tight with the set screws, to prevent it from slipping.

SETTING SLIDE VALVE OF REVERSING ENGINES.

The Link Reverse Being Used.

First, see that all the lost motion in the connecting rod, pillow block bearings and crosshead is taken up; then throw the reverse lever in last notch in quadrant, which would allow the engine to run "over"; then, remove the steam chest cover. Next, loosen eccentric (the eccentric rod of which is in direct line with the valve rod), turn the eccentric completely around, and watch the valve to see whether it laps the steam ports exactly the same amount at both ends, or travels an equal distance from its central position both ways. If it does, the valve is in proper position upon the rod. If it does not travel equally, the valve must be made to do so by adjusting it upon the valve rod, which is done by lengthening or shortening the rod, by use of jam-nuts, with which the rod is usually furnished. After this is done accurately, place the engine upon its forward center. This is done by turning the engine forward until the cross-head is about ½ inch from end of its stroke and mark the position of the cross-head upon the guide; then, with the use of long tram or dividers, mark from any convenient point on the frame to the band wheel or disc, and mark both points with prick-punch. Again turn the engine forward until the cross-head completes its stroke and comes back to the mark made on the guide; then, with the same long tram or dividers, mark the band wheel or disc as before from the prick mark already made on the frame.

Midway between punch marks on band wheel or disc, which can be found by use of dividers, will give the point which will place the engine on the "center" by turning engine back far enough to allow the long tram or divider to fit in punch mark on frame, and center punch mark on band wheel or disc. Now turn the eccentric over in the direction in which the engine is to run, until the valve gives the proper amount of lead on the front of steam port, which is about $1/32$ of an inch, and fasten eccentric with set-screw. Then turn engine over the way it is to run, and place it upon its back center. This is done exactly as before described for forward center, and if the valve is properly proportioned, it will give the required $1/32$ of an inch lead on the back steam port, and valve will be properly set for running in this direction.

Now throw the reverse lever in the last notch in opposite end of quadrant, which would allow the engine to run "under", then loosen the other eccentric (the eccentric rod of which is in a direct line with the valve rod), and proceed same as described when engine is running "over", to get valve to lap both steam ports equally. Then place engine upon either center and move the eccentric in opposite direction from the other eccentric until the valve gives the $1/32$ of an inch lead to steam port, fasten the eccentric and place the engine upon opposite center, and the amount of lead should be the same on both steam ports, and the valve properly set.

The valve being set, replace the steam chest cover, and secure the eccentrics perfectly tight by screwing the set-screws up hard to prevent them from slipping.

Always set the valve so as to run the engine backwards or "under" first.

TO SET VALVES OF DUPLEX PUMP.

Set the pistons at mid-stroke, and set the valves which are worked from the opposite side at mid-stroke also, and it will be right at all other points. The mid position of the valve can be obtained by moving it back and forth the amount of its lost motion, and dividing it so that the lead or opening on both sides will be the same.

ASCENDING HILLS.

Q. How do you ascend a hill with a traction engine?

A. When approaching a hill which you have to climb with a traction engine, see that about two inches of water shows in the glass gauge when engine is on a level. Open draft door wide, stir the fire and get it to burn briskly, and get up a good head of steam. Put reverse lever in last notch, then open throttle gradually, allowing just the necessary amount of steam to pass into the cylinder to keep the engine pulling steadily up the hill. Always start up the hill slowly; do not attempt to go up a hill at full speed, but go slowly and steadily, keeping the speed as uniform as possible by opening or closing the throttle as the case may be. Never attempt to go up a hill on a decreasing steam pressure, as there is a great liability to become stalled, in which case great damage may be done to the front end of flues. Always start on a rising steam pressure; then you know the boiler is making steam, which assures a steady ascent, as at every exhaust of the engine on an increasing or steady pressure the power becomes stronger, while at every exhaust on a decreasing pressure the power becomes weaker. Keep a uniform supply of water in the boiler at all times by use of either the pump or injector. Always when going up hill keep the draft door wide open until the steam gauge indicates that the pressure has risen almost to the blow-off point, then close the damper. Never allow steam to blow off when going up hill, as it will cause the water to raise and be carried over into the cylinder, greatly diminishing the power of the engine. Do not under any circumstances allow your engine to be stopped when going either up or down hill, as great damage may be done to the boiler.

FRICK TRACTION ENGINE.

The illustration of the left side of the Frick Traction given on opposite page represents it as being a Center Crank, Rear Geared traction engine.

This engine is constructed with an overhanging cylinder, bolted to the cast iron engine frame, which contains the locomotive style guides and both pillow block bearings for the crank shaft. It has a Cross-head Pump connected with long heater, has a specially designed Reverse Gear, also a Friction Clutch attached to the band wheel.

The round bottom fire box boiler swings in a channel iron frame, which reaches from front axle to the rear of the boiler, around the fire box, to which it is attached, and has a spring in front end only.

The wheels are made entirely of iron, with forged spokes and wrought tire, with high mud grouters bolted on.

On the channel iron frame in front of fire box is placed a heavy plank, to which two large water tanks are attached on either side. The steering wheel and band wheel are on opposite sides of engine and it has all the necessary fittings, so that with proper handling it will be found perfectly safe and reliable. The platform in rear is also supported upon the channel iron frame.

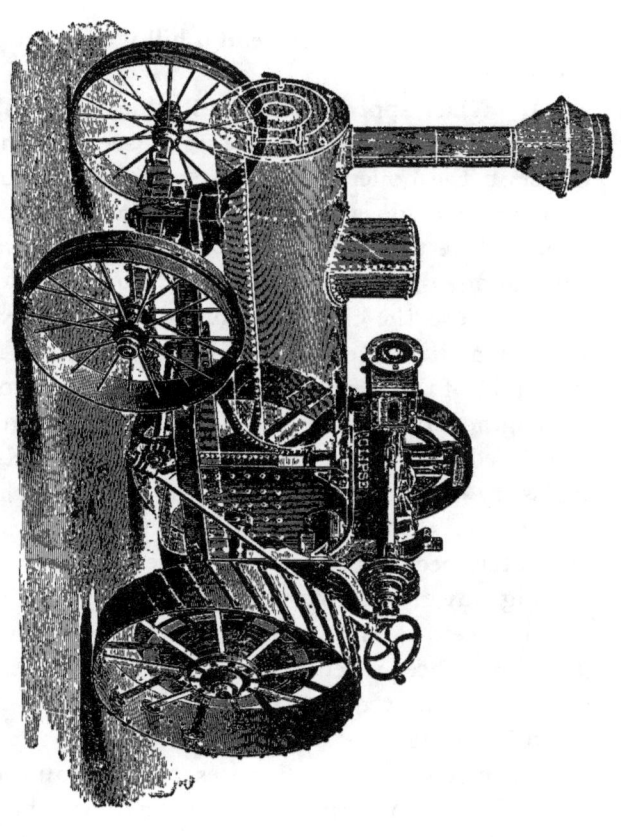

Frick Traction Engine.

DESCENDING HILLS.

Q. How do you descend a hill with a traction engine?

A. When approaching a hill which you wish to descend with a traction engine, see that the water in the boiler is at the regular height, or two inches in glass gauge when engine is on the level. Close the draught door to ash pan and open damper in smoke box when about to descend. Close the throttle almost tight, allowing just a little steam to enter cylinder, then take hold of the reversing lever, and the speed of the engine can be governed so as to descend at any speed desired, or be stopped if absolutely necessary by throwing the lever into last notch.

Do not stop while going down hill unless absolutely necessary, as there is great danger of melting out the fusible plug and damaging the crown sheet, as the water is much lower over the crown sheet when going down hill with the required amount of water in the boiler than at any other time.

When it is absolutely necessary to stop the engine for a short time when descending a hill, do not turn on the pump or injector; but open the fire door, allowing the cold air to pass over the fire, which will protect the crown sheet. This should never be done, though, except in extreme cases.

ENGINE STALLED.

There is no standard rule by which a traction engine can be gotten out of bad places upon the road, as one rule would not apply to all situations, though a few suggestions on this subject may be beneficial.

When the engine is in a bad mud hole, or on a very sandy road, and the driving wheels will not take hold, but simply turn around, the best way to get out of the fix is to hitch a good team of horses to it and pull it out with what assistance the engine can give. Though in many cases, if a quantity of straw, stones or brush, as may be most handy, is placed under the driving wheels and the power applied, the wheels may get a footing and the engine will come out all right. Old boards or rails placed in the same manner will produce the the same result.

It should be understood, however, that if the driving wheels do not take hold, but simply turn around in the sand or mud, the engine should be stopped at once and some of the aforesaid remedies tried. If you continue to work the engine under such circumstances, it will become more difficult to get out of the mire, as the wheels sink deeper in at every revolution.

CROSSING BRIDGES AND CULVERTS.

Before crossing a bridge or culvert with a traction engine, examine the stringers and floor carefully to ascertain whether it is in condition to hold the engine or not. If it appears a little weak, by laying heavy plank across for the traction wheels to run on, it may be crossed in safety, though the crossing of small bridges and culverts must be done with judgment, to prevent accident and delay.

FOAMING.

Foaming is the violent agitation of the volume of water in the boiler; it occurs only in dirty boilers and where dirty feed water is used, which causes the water to become saturated with foreign matter, such as lime, sediment, mud, oil or grease, etc.

The steam trying to escape through the scum formed by these impurities, raises the whole mass from the surface of the water in large bubbles, and causes a general frothing or foaming condition of the water level, which is indicated by the dirty appearance of the gauge cocks and joints and the cutting of piston rod and cylinder by the gritty matter carried over by the steam. Foaming does not result in carrying over so much water, but a foamy boiler does not produce as dry steam as one that is kept perfectly clean.

Q. How do you prevent a boiler from foaming?

A. To remedy foaming, or prevent it, requires frequent blowing off from the surface of the water the scum which causes it, and the use of pure feed water. It may be stopped for a while by closing the throttle valve for an instant, to give the water and scum a chance to settle.

PRIMING.

Priming in a steam boiler is the carrying over of large quantities of water by the steam to the engine, and may occur in a perfectly clean boiler. It is usually caused by too great a demand on the capacity of the boiler, too sudden and fierce firing, or after steam pressure is lowering, an increasing demand is made for it. It is usually a radical defect in the construction or capacity of the boiler, and is most frequently the result of insufficient steam space, small evaporating capacity, and lack of good circulation.

Priming is indicated by the water rising and lowering in the glass gauge more or less violently, by the clicking sound in the steam cylinder of engine as the piston forces the water from end to end, and by the regular shower of water falling from the exhaust.

Q. How do you remedy priming?

A. There are several remedies for priming, none of which will apply effectually in all cases. Where insufficient capacity is the cause, the only remedy is a larger boiler. In others, it may be prevented by carrying the water level lower, if same can be done with safety, or taking steam from side of dome instead of top, or increasing the size of steam pipe, or taking out the top row of flues, and in boilers that have no steam dome, a long dry pipe with perforated top may be of benefit.

Q. What are other causes and remedies for priming?

A. The piston rings may leak badly. If they do, they should be replaced, or made to fit cylinder perfectly. The cylinder may be badly cut by the rings. If it is, it should be re-bored and new piston rings put in.

The slide valve may be cut, and leak. In this case the valve will need re-planing and scraping, also the valve seat. If the valve is not properly set, it may also cause priming. The exhaust nozzle may be clogged with burnt oil and sediment; if it is, clean it out thoroughly.

FIRING WITH WOOD.

Always keep a level fire. Fill every open space as fast as the wood burns out. Allow as little cold air to pass through the fire as possible. *Never stir a wood fire.* Fire quickly, and keep the door shut as much as possible.

FIRING WITH STRAW.

To start the fire, push a small forkful of dry straw into funnel in fire door, leaving the small end of funnel pressed full; then touch the match to it. Begin at once to push in the straw regularly, a small quantity at a time, being very careful not to clog the main flue, and allowing ample time for straw to burn.

The fire should be raked down frequently, as the burned straw leaves a charred mass over the grates. This should be done when the funnel is full of straw, thus allowing no cold air to pass through the funnel into the main flue. Clean out the ash pan frequently, so that the natural draught may not be checked. Do not open blower until gauge shows ten or fifteen pounds steam pressure.

After steam is raised to the necessary pressure, the feeding should be regular, using small forkfuls of straw, keeping the funnel full all the time, and raking down at short intervals. Use as dry straw as it is possible to obtain.

The above will apply to any style or make of straw-burning engines.

FIRING WITH COAL.

After fire is well started with wood, throw coal into the center of grate, and do not disturb it until it is well ignited and burning briskly; then break the fire down and put in a shovel or two of coal, and so continue keeping the grates covered with a thin layer.

Always aim to put in fresh coal on a rising head of steam pressure. Never pile coal against the flue sheet or keep the fire box too full. Nothing is gained by the latter, but much is lost.

Q. Which is the more economical to burn, wet or dry coal?

A. Dry. If your coal is wet, you simply have to evaporate that much more water, which goes out of the stack instead of to the engine.

Q. How much water will one pound of coal evaporate?

A. One pound of coal will, under very favorable circumstances, evaporate twelve pounds of water, but the average evaporative power of anthracite coal is $9\frac{1}{2}$ pounds of water, and semi-bituminous coal is $9\frac{9}{10}$ pounds.

Q. If cold air is allowed to strike the flue sheet and flues, what is the result?

A. It will eventually cause them to leak.

Q. How should a fire be regulated in case of temporary stoppage by accident or otherwise under full head of steam?

A. Close the damper and keep the fire door closed; then open small door in smoke-box or the damper in chimney.

Q. Why not leave the fire door open?

A. Because it would allow the cold air to come in contact with flue sheet and flues, and consequent damage to boiler.

BANKING FIRES.

To bank a fire in a furnace, push the fire in a heap at the back of the furnace against the flue sheet; leaving a large portion of the grate open, to allow the air caused by the natural draught to pass up over the fire to the flues; then cover it over with fine coal or a layer of dry ashes, and see that the draught door is closed to prevent draught as much as possible.

This being done, the fire will last over night, and when ready to start again in the morning, all that is necessary to do is to rake the fire over the grates, open the damper and apply more fuel.

Q. What benefit is derived from banking the fire?

A. By banking the fire, the water in the boiler is kept warm over night and steam is raised quickly in the morning, saving time and fuel.

Q. When leaving a banked fire, is it practicable to shut the water out of the glass or water column?

A. Yes. In freezing weather, this may be done by closing the valve at the top and bottom of the glass; and open pet cock beneath. Care should be taken, however, to open them before the fire is started in the morning.

LAYING UP A TRACTION ENGINE.

Q. How should you prepare your engine and boiler for laying up through the winter, to protect them from frost and injury?

A. While steam is on, clean your boiler and engine thoroughly outside, scrape off all oil, grease and scale; after which apply a good coat of asphaltum paint to the boiler and smokestack. If no paint can be had, lamp black and linseed oil will answer. If this cannot be had, take rags, saturate them with grease or oil, and go over them with that.

Now open the blow-off valve, and blow the water off with a low pressure of steam, after which take out all the hand-hole plates and wash the boiler out thoroughly, removing all the mud and scale; then replace the hand-hole plates, close the blow-off valve and fill the boiler nearly full of water, after which pour in a gallon of black oil upon the water.

After this is done, open the blow-off valve again and allow the water to run out. The oil will follow the water down and cover the whole inside of boiler with a coating of oil, making as good a protection against rust as can be found.

Next, remove all the brass fittings, such as lubricator, steam gauge, safety valve, injector, check valves, pump valves, gauge cocks, water gauge, etc., etc.

Disconnect all pipes where water may lodge, in order to prevent freezing. Every pipe and valve allowed to freeze will surely burst. Unscrew all stuffing boxes and remove the packing; for unless this is done, another season you will find parts badly rusted where the packing was allowed to remain.

Take off all cylinder cocks, pet cocks, etc., from the heater and pump. All fittings should be carefully packed and laid away. Clean the flues and fire box, also the ash pan, and do not neglect to paint the ash pan both inside and outside.

Remove the back cylinder head, roll the engine forward and smear the inside of cylinder with tallow, or oil if no tallow can be had. Place the head back again and smear all the bright work, such as piston rod, connecting rod, etc., with grease. Do not forget to cover top of smoke-stack, to keep out water and snow.

If the foregoing directions are followed carefully you will find another season that your engine will be clean, free from rust and ready to serve you faithfully without any trouble or delay in starting, either in time or expense.

BELTING.

Do not tax belts by overloading. Keep them free from accumulation of dust, grease and all animal oils, as these are injurious to both rubber and leather belts.

Special care should be taken to protect the edges of rubber belts from all animal oils, as they are liable to rot the belt.

Always run the grain (or hair) side of leather belts on the pulley, as it gives greater driving power, hugs the pulley closer, is less liable to slip, and will drive 30 per cent. more than the flesh side.

Rubber belts will be greatly improved and their life prolonged, by putting on with a brush, and letting it dry, the following mixture:

Equal parts of black lead and litharge mixed with boiled oil; add enough Japan to dry it quickly. In case the rubber peels off, the same mixture can be used.

In comparison to leather belts, 4-ply rubber is equivalent to a single leather belt and 6-ply rubber to double leather belt.

To find the length of a belt, add the diameter of the two pulleys together, divide the result by 2 and multiply the quotient by 3½; then add to this product twice the distance between the centers of shafts.

When piecing a belt when pulleys are changed, multiply the difference of the diameters of the pulleys by 1½, the product will be the length of the piece required.

The seam side of rubber belt should always be placed outside and not next to pulley. In case the belt slips, coat the side next to pulley with boiled linseed oil or soap.

In lacing a belt, begin in the center and be careful to keep both ends exactly in line. Lace both ends equally tight and do not cross the lace on the pulley side of belt. Great care should be taken that the ends butting together be cut perfectly square; if not, the belt will stretch more on one side than the other, which greatly impairs its worth.

Q. What is the practical limit of belt speed?

A. Belts should not be run much over 5000 feet per minute.

Q. How then is the capacity of a belt affected by its speed?

A. It varies directly as the speed. A given belt will transmit twice the horse power if its speed is doubled within limits.

Q. Is the capacity of a belt affected by its width?

A. Yes, the capacity varies directly as the width. If a two inch belt will transmit one horse power, two such belts will transmit two horse power; and this is true whether they are run separately or joined into a four inch belt.

To preserve cotton or Gandy belting, apply with a brush a little common paint to pulley side of belt while running, to be followed shortly afterwards by a little soft oil or grease to preserve its flexibility.

If the edges of the belt become frayed from the use of belt guides or forks, the loose threads may be cut off without injury to the belt.

If the belt slips at first, consequent to the surface being ruffled by unrolling, apply a little grease, oil or soap to the pulley side to make it grip.

Armington & Sims High Speed Engine.

ARMINGTON & SIMS HIGH SPEED ENGINE.

The cylinder and steam chest of this engine are cast in one piece and bolted securely to the engine frame, which forms the front cylinder head. The cylinder is lagged with mineral wool and jacketed to prevent radiation, and it is overhanging and self-lining.

The valve is a hollow piston valve, the body of which is steel tubing with cast iron ends. It receives its motion from the shaft governor, attached to one of the band wheels, which regulates the cut-off automatically according to the variation of load. The steam is exhausted at each end of the valve by very direct passages which quickly free the cylinder, preventing back pressure.

The engine frame is cast heavy and rigid, and contains the locomotive guides for cross-head and the pillow block bearings for the crank shaft.

The double disc center crank shafts allow of two small heavy band wheels being used.

The base of this engine is cast in one piece, to which the engine frame is securely bolted, and with this arrangement, the engine needs no expensive foundation.

The engine is simple and self-contained, ranging in sizes from 11 to 450 horse-power, and is intended to run at the very high speed of from two hundred to three hundred and fifty revolutions per minute according to size, and is used extensively in driving electric lighting machinery, and where high speed and continuous work is desired.

GENERAL INFORMATION.

Never condemn an engine that is entirely new to you because it does not start off at your first effort. Study all the directions furnished by the maker. Perhaps you have overlooked some points that are of more importance than you imagine.

The above will apply to other machinery as well as engines.

When starting a new engine be sure that everything is in readiness. Turn it over by hand to see that all the revolving and reciprocating parts run freely. Start it very slowly under steam pressure and apply plenty of good oil. After it has run a short time and everything is working properly turn on more steam and continue to do so until the engine is running at its rated speed. To start it at full speed under steam pressure may result in great damage or totally destroy the engine.

An accurate machine which is thoroughly reliable is necessarily costly, but is of more value than another which merely serves a purpose.

Engineers or firemen in charge of a steam boiler should blow out the water gauge and gauge cocks every morning in order to remove the soft mud which settles in them at night when the boiler is at rest. If this is neglected, the soft mud may become baked in them which might lead to disastrous results.

Every steam boiler for whatever purpose employed, should be opened, cleaned, thoroughly examined and tested at least every six months, and with muddy feed water once a week would not be too often.

By blowing out the gauge cocks regularly you not only ascertain the height of the water in the boiler, but it prevents them from becoming choked with sediment or mud.

Do not allow the gauge cocks, glass water gauge or steam gauge to become filthy, as it shows lack of care, and furnishes evidence that the engineer who is not particular in this part of his duty is not reliable in others of equal or more importance.

Upon entering the boiler room in the morning an engineer or fireman should always ascertain whether the valves or cocks which connect the water gauge with the boiler are open or shut, otherwise he may be deceived by the appearance of the water in the tube. This precaution should never be neglected.

If an engineer or fireman discovers that there is too much water in the boiler he should blow it down to the proper level, but in doing so he must exercise judgment, vigilance and care, especially if there is a fire in the furnace.

Never allow the gauge cocks to leak at all when it is practicable to repair them, for the longer they leak the more difficult they are to repair, as under the escape of water or steam the metal wastes rapidly.

An engineer or fireman should often remove the ashes from under the boiler, or from ash pan; if allowed to accumulate, they retard the draft and interfere with combustion, thereby causing waste of fuel and interfere with the evaporating efficiency of the boiler. Also keep grates clear of clinkers; for if allowed to accumulate, they produce the same result.

Should it become necessary to blow down the water at intervals, the engineer or fireman should stand by the blow-off cock and not allow his attention to be diverted to anything else, as in a very short space of time the water may become so low as to induce stoppage or endanger the safety of the boiler.

Engineers should always be cautious when they stop or start an engine with a heavy pressure of steam in the boiler, as the vent given to the steam when starting, and the check it receives when stopping, may exert such a pressure as to strain, crack, or rupture the boiler.

The drip cocks in the cylinder should be left open when the engine is standing still, and they should not be closed until after the engine has been started and made several strokes or revolutions.

Do not open the throttle valve to its full extent in starting after the engine has been standing over night, as the quantity of steam condensed by being brought in contact with the cold pipe (particularly if it is a long one) may result in breaking the follower plate, springing the piston rod, or knocking out the cylinder head.

After opening the gauge cocks to ascertain the height of water in the boiler, they should be closed tightly to prevent leakage.

It may have been discovered that when gauge cocks are closed after being blown out, they leak badly; this is often due to the fact that mud or sand has become attached to the seat of the valve. The easiest way to remedy this difficulty is to open the cocks and let them blow out for some time, as the friction of the water in its escape will in all probability remove the obstacle.

Glass water gauges may be cleansed by removing the glass; then tying a piece of cotton waste or lamp wicking to a splint of wood, applying soap or acetic acid, and passing it through the inside of the tube; then replace the glass, and when steam is raised close the lower valve, open the drip cock, and the steam blowing through will wash the glass perfectly.

To cut a glass gauge tube.—If a glass gauge is too long, take a three-cornered file and wet it, hold the tube in the left hand with the thumb and fore-finger at the place where you wish to cut, saw it quickly and lightly two or three times with the edge of the file, and it will mark the glass. Now, take the tube in both hands, both thumbs being on opposite sides of the mark and about an inch apart, then try to bend the glass, using your thumbs as fulcrums and it will break at the mark which has weakened the tube.

Never touch the inside of the water gauge glass with iron or wire, as while the glass may be cut on the outside with a file, the slightest touch of steel or iron on the inside will cause an abrasion, the result of which is that the glass will crack and become useless.

Water gauge glasses frequently break because the steam and water connections are not in line, because the stuffing boxes are screwed down too tight, and sometimes in cold weather when struck by a cold draught of air admitted through an open door or window.

An engineer or fireman should never fill a boiler with cold water while the boiler is hot, as the injurious effect produced by contraction is similar to that produced by blowing out at a high pressure, and if persisted in will result in permanent injury to the boiler.

Exhaust steam will heat water to 212° Fahr. under atmospheric pressure.

Ten degrees extra heat in feed water means one per cent. saving in fuel.

Before blowing out the boiler the engineer or fireman should remove all the fire from the furnace, as a small quanity left in the corners, or attached to the bridge wall, might spring a seam or cause a plate to bulge.

Every engineer should know that unequal expansion and contraction is one of the evils which limit the longevity and endanger the safety of all classes of steam boilers; consequently the blowing out, the refilling, the starting of fires and the regulation of the draught should be done with judgment.

It is not necessary to fill a boiler with cold water above the second gauge cock, as the water expands under the process of the formation of steam and it will be found that there is a sufficiency of water in the boiler when steam is raised.

Single riveted seams are equal to 56% of the original strength of the sheet; double riveted seams are equal to 70%, and triple riveted seams are equal to 85%. Triple riveted seams, however, are very seldom used unless for some special purpose, as they are too heavy and thick, and would burn out rapidly if exposed to fire.

In making calculations on the strength of boilers, the factor 56 should be employed instead of 100, as 44% of the strength of the plate is lost by punching the holes for the rivets.

It should be understood that machine riveted seams in steam boilers are superior to hand made seams, as the machine thoroughly upsets the rivet and brings the two sheets in such close contact as to produce friction between the sheets at the lap, which of itself is an element of strength.

Boilers do not improve by standing idle; they will rust very rapidly.

Never use sharp chisels to cut the scale from boiler plate, as the cutting of the plate does more harm than good. Use only a light hammer.

In patching a boiler be careful not to make a pocket in which sediment may collect to cause another injury to the sheet and never put a steel patch upon an iron boiler as the two metals expanding unequally will induce trouble.

Never forget to allow for expansion when running long lines of steam pipe, whether for heating or power, as the neglect of this precaution leads to the formation of immense crooks or bends in the line of pipe wonderful to behold. There must be a slip joint somewhere in long lines of steam piping, unless expansion is allowed for.

Valves stick on their seats because they are frequently shut when cold, and when heated by the steam the valve stem becomes lengthened, and presses the valve hard into the seat.

180 YOUNG ENGINEER'S GUIDE.

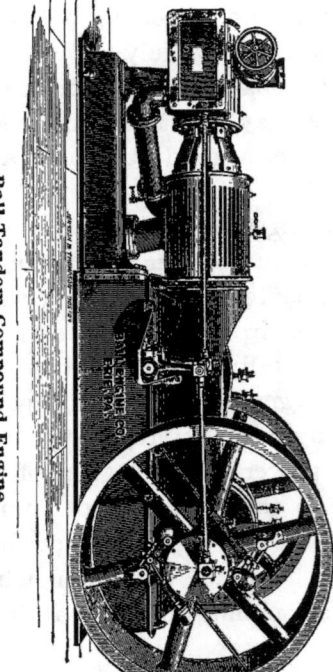

Ball Tandem Compound Engine.

BALL TANDEM COMPOUND ENGINE.

The heavy case iron base is cast in two sections, the rear part being securely bolted to the front section. To the front section is also bolted the main engine frame. This frame contains the bar guides for the cross-head, and pillow block bearing for the double disc crank shaft, and also forms the front head for the low pressure cylinder, which is securely bolted to it.

The high pressure cylinder is attached to the low pressure cylinder by two brackets securely bolted, and is supported by a pedestal, bolted to the rear part of base. By this arrangement, both the high and low pressure pistons are upon the same piston rod, which necessitates of but one cross-head, connecting rod and crank.

The valve of the high pressure cylinder is operated and completely under the control of the automatic shaft governor attached to one of the band wheels, while the valve of the low pressure cylinder receives its motion from a single eccentric on crank shaft at opposite side of engine.

When engine is running, the steam enters the high pressure cylinder first, and after performing its work there, exhausts through the receiver pipe into the low pressure cylinder, and there exerts its minimum force by expansion, and passes out to the condenser, if used, or exhausts into the open air.

Slide valves should be fitted to their seats by filing and scraping, and never by the use of emery and oil. The piston rod and valve rod may be packed with braids of hemp or cotton wicking, with rings cut from patent packing of various kinds or metallic packing.

To clean brass articles with acid is a great mistake, as with such treatment they very soon become dull. Sweet oil and putty powder followed by soap and water, is one of the best mediums for brightening brass and copper.

To frost brass work and give it an ornamental finish, boil the article in caustic potash, rinse in clean water and dip in nitric acid until all oxide is removed; then wash quickly, dry in box-wood sawdust, and lacquer while warm.

The best material for grinding in valves and stop cocks is pulverized glass. It is superior to emery for this purpose. Fine sand may be used.

To remedy a leaky angle, check or globe valve, it should be taken apart, and the valves ground to fit their seats properly with either fine sand, pulverized glass or emery.

A lever stuck between the spokes of the fly wheel of an engine for the purpose of starting it, is a very dangerous instrument, it is liable to

YOUNG ENGINEER'S GUIDE.   183

get caught and do a great amount of damage. If a lever is to be used, be sure that the steam is first turned off.

A cubic inch of water evaporated under ordinary atmospheric pressure is converted into one cubic foot of steam (approximately).

Steam at atmospheric pressure flows into a vacuum at the rate of about 1550 feet per second, and into the atmosphere at the rate of 650 feet. per second.

Condensing engines require from 20 to 30 gallons of water to condense the steam represented by every gallon of water evaporated—approximately; for most engines we say from 1 to 1½ gallons per minute per indicated horse power. Jet condensers do not require quite as much water for condensing as surface condensers. Surface condensers require about 2 square feet of tube (cooling) surface per horse power of steam engine.

The best designed boilers well set, with good draught and skillful firing, will evaporate from 7 to 10 lbs. of water per pound of first-class coal. The average result is from 25 to 60 per cent. below this.

When you have your boiler furnace to repair,

and cannot get fire clay, take common earth mixed with water, in which you have dissolved a little salt; use same as fire clay, and your furnace will last fully as long.

To make iron take bright polish like steel, pulverize and dissolve the following articles in one quart of hot water: Blue vitriol 1 oz., borax 1 oz., prussiate of potash 1 oz., charcoal 1 oz., salt ½ pt.; then add one gallon of linseed oil, mix well, bring your iron or steel to the proper heat, and cool in the solution.

To write inscriptions on metal, take 4 oz. of nitric acid and 1 oz. of muriatic acid, mix and shake well together, then cover your metal surface to be engraved, with bees-wax or soap, write your inscription plainly in the wax clear to the metal, then apply the mixed acids, carefully filling each letter. Let it remain from three to five minutes according to appearance desired, then throw on water, which stops the etching process, scrape off the bees-wax or soap, and the inscription is complete.

To remove rust from steel.—Brush the rusted steel with a paste composed of ½ oz. cyanide potassium, ½ oz. castile soap, 1 oz. whiting, and enough water to make a paste; then wash the steel in a solution of ½ oz. cyanide potassium and 2 oz. of water.

A solvent for rust.—It is often very difficult, and sometimes impossible, to remove rust from articles made of iron. Those which are most thickly coated are most easily cleaned by being immersed in, or saturated with, a solution of chloride of tin. The length of time they should remain in this bath is determined by the thickness of the rust, generally twelve to twenty-four hours is long enough. The solution ought not to contain a great excess of acid if the iron itself be not attacked. On taking them from the bath, the articles are rinsed first in water, then in ammonia, and quickly dried. The iron when thus treated has the appearance of dull silver; a simple polishing gives it its normal appearance.

One of the best varnishes for smoke stacks or steam pipes is good asphaltum dissolved in oil of turpentine.

Iron or steel immersed warm in a solution of carbonate of soda (washing soda) for a few minutes will not rust.

Cement to fasten iron to stone.—Take 10 parts of fine iron filings, 30 parts of plaster of Paris, and ½ part of sal ammoniac; mix with weak vinegar to a fluid paste and apply at once.

Cement for joints.—Paris white, ground, 4 lbs.; litharge, ground, 10 lbs.; yellow ochre, fine, ½ lb.; ½ oz. of hemp, cut short; mix well together with linseed oil to a stiff putty. This cement is good for joints on steam or water pipes; it will set under water.

The average consumption of coal for steam boilers is 12 pounds per hour for each square foot of grate surface.

One ton of coal is equivalent to two cords of wood for steam purposes.

Doubling the diameter of a pipe increases its capacity four times.

A cubic foot of water contains 7½ gallons.

A gallon weighs 8⅓ pounds.

Water expands ⅑ of its bulk in freezing.

Ice weighs 56½ pounds per cubic foot.

Engineers can judge of the condition of their machinery by the tone it gives out while running. Every make of engine has a peculiar tone of its own. The engineer becomes accustomed to that, any any departure from it at once excites a suspicion that all is not right. The engineer may not know what is the matter, he may have no ear for music, but the change in tone of his machine will be instantly preceptible and will start him upon an immediate investigation.

An Indicator is an instrument used to determine the indicated horse power of an engine; it shows the action of the steam in the cylinder and serves as a guide in setting valves to get the greatest amount of energy from the steam used.

Atmospheric pressure is the weight of the air.

To take lime from injector tubes, mix one part muriatic acid and ten parts soft water. Immerse tube in this mixture over night.

Compound for Cooling Heavy Bearings.— For cooling heavy pillow block bearings, or the steps of upright shafts, the following will be found very valuable: Four pounds of tallow, one-half pound of sugar of lead, three-fourths pound plumbago. When the tallow is melted (not boiling) add sugar of lead and let it dissolve; then put in the plumbago, and stir the whole mass until cold.

A mixture of soft soap and black lead makes an excellent lubricant for gears, as it lessens the abrasion and noise and has the advantage over tallow of not becoming hard. It is also easily removed should it become necessary to clean the parts on which it has been used.

The axles and axle arms of a traction engine should be well greased or oiled before moving, to prevent them from being cut and wearing both hub and axle rapidly.

WORKSHOP RECIPES.

LOAM.—Mixture of brick, clay and old foundry sand.

PARTING SAND.—Burnt sand scraped from the surface of castings.

BLACK WASH.-Charcoal, plumbago and size.

BLACKENING FOR MOLDS.—Charcoal powder, or in some instances fine coal dust.

MIXTURE FOR WELDING STEEL.—One part sal ammoniac, ten parts borax, pounded together and fused until clear. Then it is poured out and after cooling, reduce to powder.

RUST-JOINT CEMENT.—(Quickly setting.) One part sal ammoniac in powder (by weight), two parts flour of sulphur, eighty parts iron borings, made into a paste with water.

RUST JOINT.—(Slowly setting.) Two parts sal ammoniac, one part flour of sulphur, 200 parts iron borings. The latter cement is the best if the joint is not required for immediate use.

RED LEAD CEMENT FOR FACE JOINTS.—One part white lead, one part red lead, mixed with linseed oil to the proper consistency.

CASE HARDENING.—Place horn, hoof, bone dust, or shreds of leather, together with the article to be case hardened, in an iron box; subject to blood red heat, then immerse the article in in cold water.

CASE HARDENING WITH PRUSSIATE OF POTASH.—Heat the article, after polishing, to a bright red; rub the surface over with prussiate of potash; allow it to cool to dull red, and immerse it in water.

CASE HARDENING MIXTURES.—Three parts of prussiate of potash, one part sal ammoniac; or, one part of prussiate of potash, two parts sal ammoniac and two parts bone dust.

GLUE TO RESIST MOISTURE.—One pound of glue, melted in two quarts of skim-milk.

MARINE GLUE.—One part of India rubber twelve parts mineral naphtha or coal tar. Heat gently, mix, and add twenty parts of powdered shellac. Pour out on a slab to cool. Heat to about 250 degrees and it is ready for use.

GLUE CEMENT TO RESIST MOISTURE.—One part glue, one part black rosin, ¼ part red ochre, mixed with the least possible quantity of water; or, four parts of glue; or, one part oxide of iron, one part of boiled oil (by weight).

BABBITTING BOXES.

When the babbitt in a box is badly worn, and needs re-babbitting, remove the cap, take out the shaft and chip all the old babbitt out of both box and cap; then replace the shaft in the box, and line it up perfectly level and square by putting liners in between the shaft and the edges of the box; then put stiff putty around the shaft and against the box at both ends, to prevent the babbitt from running out; then heat the babbitt metal until it runs freely, and pour it into the box until it is full; then put on the cap, and place about the same amount of liners between its ends and the top of the shaft as was put under the shaft, with long liners of sheet iron or tin extending from one end to the other of the box, parallel with and on both sides of the shaft; then put putty around the shaft and against the cap at both ends; heat the metal again, and pour it in through the oil hole. After it is cool, remove the cap and liners, drill out the oil hole and replace the cap, being careful to put just enough liners under it so that the box will be tight and still have the shaft run cool.

COMPOUND ENGINES.

The Compound Engine dates from the year 1781, when Hornblower, a contemporary of Watt, conceived the idea of utilizing the force in the exhaust steam of the simple engine in a second cylinder.

From his crude design, the constant progress of experiment has developed the marvelous engines now used in ocean steamers, and in both large and small power plants, also on locomotives. Some of the compound engines built in the early part of the century show results, according to the records, not far behind the best attainable in modern times.

The era of the Compound locomotive engine began in Europe in 1876, but in this country half a dozen years would almost cover its history.

However, in this short time, its advantages in putting to profitable use the entire force of the steam supplied, has been so clearly shown, that it has evidently come to stay. Its availability as an efficient, economical, powerful high speed locomotive, demonstrates the value of the Compound as a farm traction engine, and makes it plain that it will be extremely serviceable on this class of engines.

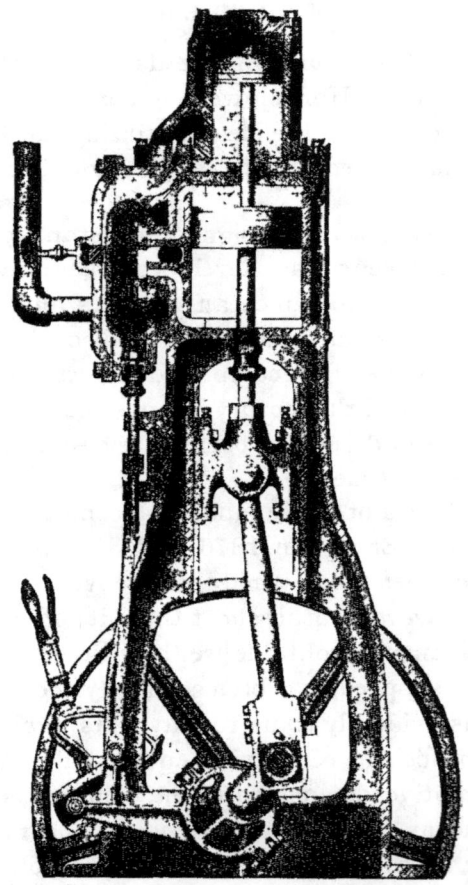

Woolf Compound Engine and Valve Gear.

The type of compound belonging to the class known as the "Woolf" is especially adapted for use on a traction and farm engine.

The cylinders set tandem, are bolted together on a matched joint so as to preserve perfect alignment, and they are brought so close together that the single cylinder head or partition between them serves for both cylinders.

Where the piston rod of the high pressure piston passes through this partition or head, durable metallic self-packing rings are provided.

Only one steam chest is required, it is located on the low pressure cylinder, and is not a receptacle for steam from the boiler, but receives the exhaust steam from the high pressure cylinder for distribution to the low pressure cylinder.

The valve seat is located in the receiver.

The valve is a single casting and performs all the functions for both high and low pressure cylinders, and instead of being loaded with high pressure steam, there is comparatively light pressure all over the back of the valve. This pressure is offset or balanced by the high pressure steam which has access to a smaller area on its face and the valve is well balanced without complications.

The cut shows a small piston placed in the steam chest cover; it is used on traction engines and forms a counter-balance to hold the valve properly to its seat when the engine is moving down hill or running with a light load.

The design has the merit of compactness, extreme simplicity in construction and renders available all the benefits obtainable by compound expansion.

Its advent marks a very important improvement in the construction of farm engines, as its efficiency does not depend on extreme high pressure steam nor on unusual skill in handling.

Q. How is pistons removed from Woolf Compound Cylinders?

A. Remove the high pressure cylinder, then loosen the check nut or piston rod against cross-head, unscrew rod from cross-head and piston rod, with both pistons and partition, can be taken out.

All parts should be replaced in the reversed order and care should be taken to tighten all nuts and bolts loosened.

SETTING WOOLF VALVE.

First take up all "lost motion" and place the engine on the forward center, by putting the wrist pin, crank pin and center of crank shaft in line. This is accomplished in precisely the same manner as when "setting slide valve of reversing engine" (see page 149): Then remove the steam chest cover, next loosen set screw in eccentric and set eccentric with its throw opposite the crank pin, so that the valve will have no motion or travel when throwing the reverse lever backward and forward to extreme ends of quadrant.

Should there be any movement whatever to the valve when throwing the reverse lever back and forth with engine on either center, the eccentric must be adjusted, by turning it on the shaft, a little at a time, either up or down as the case requires, until there is no motion to the valve, then fasten eccentric securely with the set screws.

This being done properly see if the valve has an equal amount of "lead" at each end, by turning the engine over in the direction it is to run and watching the "lead" openings when crank reaches center. If it has an equal amount of "lead" on each steam port the valve is properly set and requires no adjustment. If the valve has more "lead" on one end than on the other move it by adjusting the valve stem towards the end having the greatest amount of "lead" a distance equal to one-half the difference in the "lead," which, if valve is properly constructed, will give the exact amount of "lead" necessary and the valve will be accurately set. Replace steam chest cover and be sure to tighten all set screws and nuts loosened.

The adjustment of the Woolf valve by the valve stem is accomplished in different ways according to the manner in which it is constructed by the several manufacturers.

Set the Woolf valve so as to run the engine forward or "over" and if the valve gear is rightly constructed the valve will also be set to run the engine backward or "under."

The setting of the Woolf valve on either "Simple" or "Compound" engines is accomplished in the same manner.

EXAMINATION OF ENGINEERS APPLYING FOR A LICENSE.

QUESTIONS WITH ANSWERS.

Q. How long have you run an engine?

Q. Have you done your own firing?

Q. What kinds of engines have you run?

Q. What would be your first duty if called upon to take charge of an engine?

A. To ascertain the exact condition of the boiler and all its attachments, such as safety valve, steam gauge, water gauge and cocks, pump, injector, blow-off valve, etc.; also the engine.

Q. How often would you blow off your boiler?

A. Once a week or month, according to the condition of feed water used.

Q. How many feet of heating surface is allowed per horse-power by builders of boilers?

A. From 12 to 15 square feet for flue and tubular boilers.

Q. How much steam pressure will be allowed on a boiler 40 inch diameter, ⅜ thick, pounds T. S., ⅙ T. S. factor of safety?

A. One-sixth of tensile strength of plate multiplied by thickness of plate, divided by one-half of the diameter of boiler gives safe working pressure.

Q. How do you estimate the strength of a boiler?

A. By its diameter and thickness of material, single or double riveted.

Q. Which is the stronger, single or double riveted?

A. Double riveted is from 14 to 18 per cent. stronger than single.

Q. What is the use of a mud drum on a boiler?

A. To collect all the sediment from the water used in the boiler.

Q. What causes sediment to accumulate in boilers?

A. The use of impure or muddy water.

Q. How often should it be blown out?

A. Three or four times a day.

Q. How much grate surface do boiler mak- allow per horse-power?

A. About two-thirds of a square foot.

Q. What is the steam dome of a boil for?

A. For dry steam to collect in.

Q. Of what use is a safety valve on a boiler?

A. To prevent overpressure.

Q. What is your duty with reference to it?

A. Open it once or twice a day to see that it is in good order.

Q. Of what use is a check valve?

A. To prevent the water in boiler from returning into the pump or injector.

Q. What effect has cold water on hot boiler plates?

A. It will fracture them.

Q. How should the gauge cocks be located on a boiler?

A. So that the lowest gauge cock is about 1½ inches above the top row of flues.

Q. Where should the blow-off valve be located?

A. At the bottom of the fire box in locomotive style of boiler, or in the mud drum when used.

Q. How would you have check valve arranged?

A. With a stop cock between the boiler and check valve.

Q. Does a man-hole in the top shell of boiler weaken it?

A. Yes, to a certain extent.

Q. How many valves in a common plunger pump?

A. Two, a receiving and a discharge valve.

Q. How are they situated?

A. One at suction, the other at discharge end.

Q. How do you find the proper size of safety valve for boiler?

A. Two square feet of grate surface is allowed for one inch area of common lever valve, or three square feet of surface to one inch area of spring valve.

Q. Why do pumps fail to work at times?

A. Leak in the suction, leak around the plunger, leaky check valve, or valve out of order.

Q. Why do injectors fail to work at times?

A. Leak in suction, grit or dirt under seat of valve, or valve not seated properly.

Q. How often should a boiler be examined and tested?

A. Twice a year at least.

Q. How would you test a boiler?

A. By tapping it with a light hammer, and hydrostatic test, using warm water.

Q. Where does the feed water enter the boiler?

A. Below the water level, where the feed water will not strike the heated plates.

Q. What causes priming of boilers?

A. Too high water, not steam space enough, dirty feed water, misconstruction of boiler, or engine being too large for its capacity.

Q. How can you keep boilers clean or remove scale from them?

A. By regularly cleaning them thoroughly, and by the use of compounds.

Q. If you found a thin plate in your boiler what would you do?

A. Patch it on the inside, first cutting out the damaged part.

Q. Why cut out the damaged part of sheet, when putting on a patch?

A. To allow the water to rest against the patch to protect it from the intense heat.

Q. What would be the result if the damaged part of sheet was not cut out?

A. The water not coming in contact with the patch, it would soon bulge from the heat and crack.

Q. Why patch it on the inside?

A. Because the action that has weakened the plate before will act upon the patch, when this is worn it can be replaced.

Q. If you found you had to put on several patches what would you do?

A. Reduce the steam pressure.

Q. If you found a blister what would you do?

A. Cut it out and put a patch on the fire side.

Q. If you found a plate buckled or sagged what would you do?

A. Put a stay bolt through the center of the sag.

Q. What would you do with a cracked plate?

A. Cut out the damaged part and put a patch over it.

Q. How would you change the water in a boiler when steam is up?

A. By supplying more feed water and opening the surface blow-off at short intervals.

Q. When blowing off a boiler, would you leave the blow-off cock to attend to other work?

A. Never.

Q. What would you do to relieve the pressure on the boiler if the safety valve was stuck and steam constantly rising?

A. Cover the fire with coal or ashes, close draught door and open damper in smoke box; work off the steam with the engine and when boiler has cooled down put the safety valve in working order.

Q. What may be the result if you allow the water in the boiler to get low?

A. Burning of the crown sheet and flues and perhaps cause an explosion.

Q. Would you turn feed water into a boiler in which the water was very low?

A. Never, without first pulling the fire or covering it with dry ashes and allowing the steam to go down.

Q. If you allow water in the boiler to get too high what would be the result?

A. It would cause priming or foaming.

Q. Is priming or foaming dangerous to an engine?

A. Yes. It may cause breaking of cylinder head and wrecking of the engine.

Q. What are other causes for foaming or priming of a boiler?

A. Dirty and impure water.

A. W. STEVENS TRACTION ENGINE.

The position of the side crank engine upon the boiler allows of having the Rear Gear traction attachment.

The Engine frame, guides for cross-head, cylinder, steam chest, saddles, brackets and both pillow block bearings for crank shaft are cast in one piece and bolted to the boiler.

The frame is cast oval, and cross-head guides are of the locomotive style.

The Engine is furnished with a Friction Clutch, a specially designed Reversing Gear, Governor, Marsh steam pump, Injector; and is mounted upon an open bottom fire box locomotive boiler, with ash pan under fire box and dome at rear end.

The boiler is mounted upon the traction wheels by brackets bolted to the rear end, which contain the boxes for the main axle and cross shaft.

The traction wheels are of the cast iron rim type, with spokes cast in both rim and hub.

The steering wheel and band wheel are on opposite sides of boiler, and both engine and boiler are supplied with all necessary fittings.

A. W. Stevens Traction Engine.

Q. How would you stop foaming?

A. Close the throttle long enough to show the true level of water. If the level of the water is sufficiently high, feeding and blowing off will usually correct the difficulty.

Q. What would you do if you discovered the water gone from sight in the water glass?

A. Pull the fire or cover it over with dry ashes, and allow the boiler to cool off as quickly as possible; and would not open or close any of the steam outlets.

Q. What is a traction engine?

A. A traction engine is an engine the power of which is transmitted to the driving or ground wheels by a combination of gearing.

Q. What is an exhaust pipe?

A. The pipe through which the exhaust steam escapes from cylinder to smoke-stack.

Q. What is the feed pipe?

A. The pipe through which the feed water passes from pump or injector to the boiler.

Q. What is the steam pipe?

A. The pipe through which steam is taken from the dome to the steam chest.

Q. What is a pet cock?

A. A small cock used in check valves, pipes

and places where draining off water is necessary to prevent freezing.

Q. What is clearance in a steam cylinder?

A. It is the space between the cylinder head and piston head when the latter is at end of the stroke.

Q. What is "cushion" in a steam cylinder?

A. Cushion is the compression of steam let in through the lead of the valve in the clearance of the cylinder, and is for the purpose of catching the weight of the piston and rod, cross-head and connecting rod when the engine reaches the end of each stroke. It also keeps the engine from pounding.

Q. How much water would you blow off at any one time while running?

A. Never blow off more than one gauge.

Q. What are your general views regarding boiler explosions?

A. The greatest causes are from ignorance, carelessness and neglect.

Q. What precaution should you take if necessary to stop with a heavy fire in the furnace?

A. Close the draught door, and put the injector or pump at work.

Q. What is the proper height to carry water in the boiler?

A. About 2½ inches above top row of flues.

Q. At what pressure should you blow off a boiler?

A. At a pressure not to exceed ten pounds.

Q. If you wished to increase the power of an engine what would you do?

A. Increase its speed or get higher steam pressure.

Q. How do you find the horse power of an engine?

A. Multiply the speed of piston travel in feet per minute by the total effective pressure upon the piston in pounds, and divide the product by 33,000.

Q. What is meant by "brass bound"?

A. Brass bound means that the half brasses touch each other and cannot be driven up any closer by the key.

Q. How would you remedy a brass bound box on crank pin or wrist pin?

A. Take off the boxes and file off the top and bottom edges, being careful not to take off too much.

Q. Does a perfect fitting or an imperfect fitting valve have the most friction?

A. An imperfect fitting one.

Q. How would you refit an imperfect fitting or leaky valve?

A. It should be re-faced on a planer or filed and scraped until it fits seat perfectly tight.

Q. How is a steam engine rated?

A. By amount of horse power developed.

Q. What is a foot-pound?

A. One pound of force exerted through one foot of space.

Q. How many foot-pounds are required to lift 100 pounds one foot?

A. One hundred.

Q. How many foot-pounds required to lift one pound 100 feet?

A. One hundred.

Q. To lift 110 pounds through 300 feet how many foot-pounds required?

A. $300 \times 110 = 33,000$ foot-pounds.

Q. Would that equal one horse power?

A. Yes, if done in one minute.

Q. Suppose it took two minutes?

A. Then there would be only half a horse power, or $33,000 \div 2 = 16,500$ foot-pounds per minute.

Q. Is it correct to say "horse power per minute" or "horse power per hour"?

A. No. When an engine is working at the rate of 10 horse power, it is doing 10 horse power all the time. It is an error to assume that such an engine is doing 10 horse power per minute, and 10 x 60 equals 600 horse power per hour. When it is said that an engine uses 20 pounds of steam per horse power per hour, it is meant that this amount of steam is used per hour for each horse power developed.

Q. How is the foot-pounds of work done by a steam engine, found?

A. Multiply the average pressure per square inch during the stroke by the number of square inches in the piston, and by the number of feet through which the piston has moved.

Q. What do you understand by the "mean effective pressure"?

A. The mean pressure is the average pressure pushing the piston through the stroke, which is about one-third the pressure in the boiler. There is generally some back pressure working against it, therefore the "effective" pressure is only the difference between the two. It can only be determined accurately by measurements from an indicator diagram.

Q. What is a single acting engine?

A. An engine in which the steam acts on one side of the piston only.

Q. How do you find the "piston's speed"?

A. On double acting engines, multiply the stroke in inches by two and by the number of revolutions per minute and divide by 12.

Q. Why multiply the stroke in inches by 2?

A. Because in double acting engines there are two working strokes to each revolution.

Q. Why do you divide by 12?

A. To reduce the inches to feet.

Q. How is the "piston's speed" of a single acting engine found?

A. Multiply the stroke in inches by the revolutions per minute and divide by 12.

Q. What is the horse power developed by an engine, say 12 x 24 inch, running 125 revolutions per minute, with 40 pounds mean effective pressure?

A. Area = 12 x 12 x .7854 = 113.0976 sq. ins.
Piston speed = 24 x 2 x 125 ÷ 12 = 500 feet per minute.
M. E. P. = 40 lbs.
Then
$$\frac{113.0976 \times 500 \times 40}{33000} = 68.544 \text{ H. P.}$$

Q. What is a single valve engine?

A. It is an engine in which a single valve controls the admission and distribution of steam to both ends of the cylinder, or a common slide valve engine.

Q. What is a four valve engine?

A. An engine which has separate steam and exhaust valves for each end both top and bottom of cylinder, such as a Corliss engine.

Q. Into what three classes are engines divided with reference to the manner in which they are governed?

A. Throttling engines, Automatic cut-off and Governor engines.

Q. What is an Automatic cut-off engine?

A. An engine in which the amount of steam supplied is automatically cut off at various points in the stroke, in accordance with the load and pressure. In Throttling engines the volume admitted is constant and the pressure varied. In Automatic cut-off engines, steam is admitted at the highest available pressure, and the volume is varied to meet the requirements of the load. In Governor engines, the steam is admitted and cut off by the governor.

Q. What is a Throttle governed engine?

A. An engine in which the amount of steam supplied is regulated by changing the pressure at which it enters the cylinder in accordance with the variation of the load.

Q. What is a Governor engine?

A. An engine in which the supply of steam is regulated by the governor.

Q. Into what classes may the Automatic cut-off engine be divided?

A. Into two classes: The four valve engine, in which the cut-off is usually effected by a detaching mechanism or trip under the control of the governor; the single valve engine, in which the point of cut-off is varied by changing the amount of travel of the valve.

Q. Give examples of the single valve type.

A. High speed, self contained engines which have shaft governors.

Q. What are their advantages?

A. High rotative speed, compactness, portability, light weight and simplicity.

Q. Are they more economical than the four valve engine?

A. No; the four valve engines are the more economical.

Q. Give a prominent example of the four valve engine.

A. The Reynold's Corliss.

Q. What is meant by an engine running "over"?

A. The top of the drive wheel running away from the cylinder.

Q. What is meant by an engine running "under"?

A. The top of the drive wheel running towards the cylinder.

Q. Which way are engines generally run?

A. "Over."

Q. What advantages do engines have in running "over"?

A. The pressure of the cross-head on engines running over, is always downward upon the guides; for when the pressure is on the head end of the piston, the pressure against the connecting rod which is pointing upward, reacts by pressing the cross-head down upon the lower guide, and when the pressure is on the crank end of the cylinder, the cross-head will be dragging the crank, and as the crank is below the center line, it will pull the cross-head down upon the lower guide, while if the engine is running under, the pressure of the cross-head will be upon the top guide, both on the outward and inward strokes, and unless the cross-head is nicely adjusted to its guides and the guides are perfectly parallel, the cross-head will be lifted when subjected to thrust, and will fall on the center by its own weight, causing the engine to pound.

Q. At what point in the stroke is the pressure on the cross-head greatest with a uniform pressure in the cylinder?

A. When the crank is at right angles to the guide.

Q. How does the relative length of the connecting rod affect this pressure?

A. The longer the connecting rod as compared with the crank, the less will be the pressure on the guides.

Q. What is the usual ratio of connecting rod to crank?

A. The connecting rod is from four to six. times the length of the crank.

Q. Are there any objections to a long connecting rod?

A. A long connecting rod makes a long engine, and makes extra cost in the bed or frame and the room occupied. The longer rod is heavier, and brings extra weight upon the cross-head, guides and crank-pin. The long rod also lacks stiffness unless excessively heavy.

Q. What determines the length of the crank?
A. The stroke.

Q. What limits the stroke?

A. The piston's speed limits the length of stroke allowable with a given rotative speed, or the number of revolutions per minute with a given stroke.

Q. What is the practical limit of piston's speed?

A. Engines of from four to six foot stroke can run at from seven to eight hundred feet piston's speed per minute. Those of shorter stroke should not run over six hundred feet.

Q. Why do high speed engines have a short stroke in comparison with the diameter of their cylinders?

A. So that they can run at a high rate of speed without exceeding the limit of piston's travel.

Q. What is the office of the fly wheel?

A. It maintains a uniformity of motion of the crank, notwithstanding the unequal moving force upon the crank-pin.

Q. Is the force upon the crank-pin unequal, even when the pressure from the cylinder is uniform throughout the stroke?

A. Yes. No matter what the pressure on the piston is, it has no effect in turning the engine when the crank is in line with the guides, which is termed "on the center." As the crank gets away from the centers, the effect of a given pressure becomes greater, and reaches its maximum when the crank is nearly at right angles with the guides.

Q. How does the fly-wheel counteract the jerky motion of the crank which would result from this?

A. By its tendency to resist an excessive moving force, and by its momentum keeps the engine in motion when the moving force is deficient.

Q. What would you do if the cylinder gets worn or cut from too tight rings or lack of oil?

A. Rebore the cylinder.

Q. What would you do if the crank-pin heats, gets worn or cut?

A. If bent it should be turned true again; if not bent it can be filed and polished perfectly true by hand.

Q. What would you do if the main bearings heat?

A. Loosen the caps and apply plenty of good oil. If this does not stop it take off the caps, examine the oil holes to ascertain why the oil does not reach the bearing. If the bearings have become rough and cut, the shaft will have to be smoothed again.

Q. Would any harm result from starting an engine with the drip cocks closed?

A. Yes, the condensed steam filling the space would smash the cylinder or piston head.

Q. What do you mean by atmospheric pressure?

A. The weight of the atmosphere, which is 14.7 lbs. per square inch at sea level.

Q. How hot can you get water with exhaust steam under atmospheric pressure?

A. 212° Fahr.

Q. Does atmospheric pressure have any influence on the boiling point?

A. It does.

Q. Would you run an engine with throttle wide open, or partly open?

A. Wide open on governor engines, as it is more economical.

Q. How many pounds of water required per horse power for the best engines?

A. From 25 to 30 pounds.

Q. At what temperature has iron the greatest tensile strength?

A. About 600 degrees.

Q. How much water is consumed (in pounds) per hour per indicated horse power?

A. From 25 to 60 pounds.

Q. How much steam will be evaporated from one cubic inch of water under atmospheric pressure?

A. About one cubic foot, approximately.

Q. How much coal is consumed per hour per indicated horse power?

A. From two to seven pounds.

Q. How much does one cubic foot of fresh water weigh?

A. 62½ pounds.

Q. How much does one cubic foot of iron weigh?

A. 486 6/10 pounds.

Q. What does one square foot of half inch boiler iron weigh?

A. Twenty pounds.

Q. For steam purposes, how much wood is required to equal one ton of coal?

A. About 4000 pounds of wood.

Q. Of what does coal consist?

A. Carbon, nitrogen, sulphur, hydrogen, oxygen and ash.

Q. What are their relative proportions?

A. There are different proportions in different specimens of coal. The average per cent is carbon eighty, nitrogen one, sulphur two, hydrogen five, oxygen seven, ash five.

Q. Of what is air composed?

A. It is composed of nitrogen and oxygen in the proportion of seventy-seven of nitrogen and 23 of oxygen.

Q. Of what does water consist?

A. Hydrogen and oxygen in the proportion of one of hydrogen to eight of oxygen by weight.

Q. What are the different kinds of heat?

A. Latent heat, sensible heat, and sometimes total heat.

Q. What is meant by latent heat?

A. Heat that does not affect the thermometer and which expends itself in changing the nature of a body, such as turning ice into water or water into steam.

Q. Under what circumstances do bodies get latent heat?

A. When they are passing from a solid to a liquid state, or from a liquid to a gaseous state.

Q. How can latent heat be recovered?

A. By bringing the body back from a state

of gas to a liquid, or from a liquid to a solid.

Q. If the power is in coal, why should we use steam?

A. Because steam has some properties which make it an invaluable agent for applying the energy of the heat to the engine.

Q. What is steam?

A. It is an invisible elastic fluid generated from water by the application of heat.

Q. What are its properties which make it so valuable to us?

A. First. The ease with which we can condense it.

Second. The small space which it occupies when condensed.

Third. Its great expansive power.

Q. What do you understand by the term "horse power"?

A. A horse power is equivalent to raising 33,000 pounds one foot per minute.

Q. What do you understand by "lead" on an engine valve?

A. Lead on a valve is the admission of steam into the cylinder before the piston completes its stroke.

Q. What are considered the greatest improvements on the stationary engine in the past forty years?

A. The Corliss valve gear, the governor, the compound and triple expansion.

Q. What is meant by triple expansion engine?

A. A triple expansion engine has three cylinders using the same steam expansively in each one.

Q. What is the clearance of an engine as the term is applied at the present time?

A. Clearance is the space between the cylinder head and the piston head with the ports included.

Q. What is the principal which distinguishes a non-condensing from a condensing engine?

A. Where no condenser is used, and the exhaust steam is open to the atmosphere, it is a non-condensing engine.

Q. Why do you condense steam?

A. To form a vacuum and thus remove the atmospheric and back pressure that would otherwise be on the piston, thereby getting more useful work out of the steam.

Q. What is meant by vacuum?

A. A space void of all pressure.

Q. How can you maintain a vacuum?

A. By the steam used being constantly condensed by the cold water or cold tubes, and the air pump constantly clearing the condenser.

Q. Why does condensing the used steam form a vacuum?

A. Because a cubic foot of steam at atmospheric pressure shrinks into about one cubic inch of water.

Q. What is a condenser as applied to an engine?

A. The condenser is that part of an engine into which the exhaust steam enters and is condensed.

Q. About how much gain is there by using the condenser?

A. Seventeen to twenty-five per cent. where cost of water is not figured.

Q. What do you understand by the use of steam expansively?

A. Where steam admitted at a certain pressure is cut off and allowed to expand to a lower pressure.

Q. How many inches of vacuum gives the best result in a condensing engine?

A. About 25 inches.

Q. What is meant by a horizontal compound tandem engine?

A. One cylinder being back of the other with two pistons on the same rod.

Q. What do you understand by lap?

A. Outside lap is that portion of the valve which extends beyond the ports when valve is placed on the center of its travel; inside lap is that portion of valve which projects over the ports on inside or toward the middle of the valve.

Q. Of what use is lap?

A. It gives expansion to the steam in the cylinder.

Q. What is the dead center of an engine?

A. The point where the center of shaft, center of wrist-pin and center of piston rod are in the same straight line.

Q. From what cause do belts have power to drive shafting?

A. By friction and cohesion.

Q. When would you oil an engine?

A. Before starting it and as often while running as is necessary.

Q. What is the tensile strength of American boiler iron?

A. 40,000 to 60,000 pounds per square inch.

Q. What are the principal defects found in boiler iron?

A. Imperfect welding, brittleness, low ductility.

Q. What is the advantage of steel as a material for boiler plate?

A. Tensile strength, ductility, homogeneity, malleability and freedom from laminations and blisters.

Q. What are the disadvantages of steel as a material for boiler plate?

A. It requires greater care in working than iron and is subject to flaws induced by the pressure of gas bubbles in the ingots from which the plates are made.

Q. How far apart should stay bolts be put in a boiler?

A. They vary from 4 to 6 inches apart, depending on thickness of plates, size of stay bolts and amount of steam pressure to be carried.

Q. Where can examination for engineer's license be had.

A. Write to the "Secretary of State" in which you live also ask for address of the state inspector located nearest to where you live.

STEAM ENGINE INDICATOR.

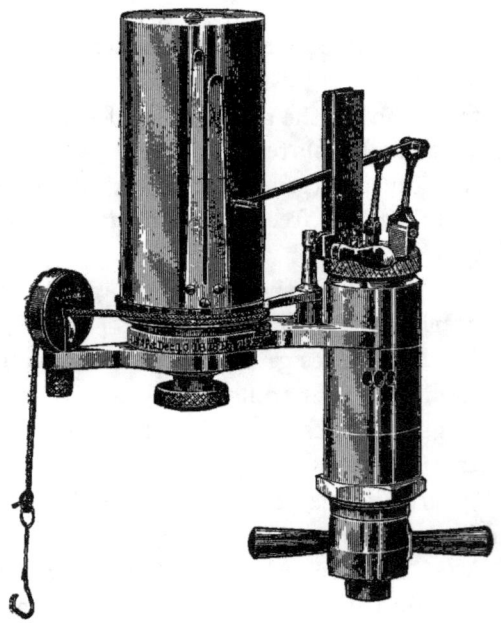

Tabor Indicator.

Having been called upon to explain the workings and use of the indicator in steam engineering practice, it might be well to state, that to give a full description embodying in detail, the construction of different makes of indicators, the mechanism, workings, manner of attaching to different makes of engines, method and rules for figuring the cards, etc.; that would enable the young inexperienced engineer to put an indicator to practical use, would require a volume as large or larger than this book; therefore it will be understood, the information given is only a brief outline on this exhaustive subject which will enable the reader to clearly understand the office, and the results obtained from the practical use of an indicator.

The young engineer who desires to put the steam engine indicator to practical tests, should seek the assistance of an expert, experienced in the application of it, until such time as he has gained the required knowledge to apply it correctly, as an indicator is a very delicate piece of mechanism and, although calculated to serve good ends, it should never be applied by unskilled hands.

It being impractical to try to explain, in a limited space, the intricate workings of the indicator, the following information tells briefly, and will give a general idea of the indicator and the results obtained from the proper use of it.

The steam engine indicator is designed to

furnish a diagram on paper of the action of the steam within an engine cylinder, from commencement to termination of piston stroke, during one or more revolutions of the crank shaft.

This is accomplished by properly attaching the indicator to a pipe or pipes leading into one or both ends of the cylinder and connecting the string which operates the paper drum, by a reducing motion, to the cross-head of the engine.

The reducing motion is necessary and must be accurately figured in order that the motion of the paper drum will coincide with motion of the cross-head on a greatly reduced scale, so that the diagram on the paper will not exceed 3½ or 4 inches in length.

As different types of engines require different kinds of reducing motions, it will be necessary for the operator to construct one that will properly attach to the engine to be indicated, in order to get a perfect card.

From this diagram or card the following data is derived and the mean effective steam pressure throughout the stroke or strokes is figured.

A. Atmospheric line.
B. Initial pressure in cylinder.
C. Piston stroke to cut-off. Automatic engine Figs. 1 and 2.
C. Reduction of pressure from commencement of stroke to cut-off, Figs. 3 and 4.
D. Piston stroke to exhaust.
E. Terminal pressure.
F. Gain in economy due to expansion Figs. 1 and 2.
G. Back pressure, if non-condensing engine, Figs. 1, 3 and 4.
H. Vacuum obtained, if condensing engine, Fig. 2.
I. Piston stroke to exhaust closure.
J. Loss produced by valve having no lead Fig. 4.
M. Height due to initial pressure.

By refering to the cuts of diagrams and comparing the letters marked thereon, with corresponding letter in table above, the points refered to in the latter will be readily understood; the action of the steam in the cylinder can be traced and the manner in which the mean-effective pressure throughout the stroke is determined.

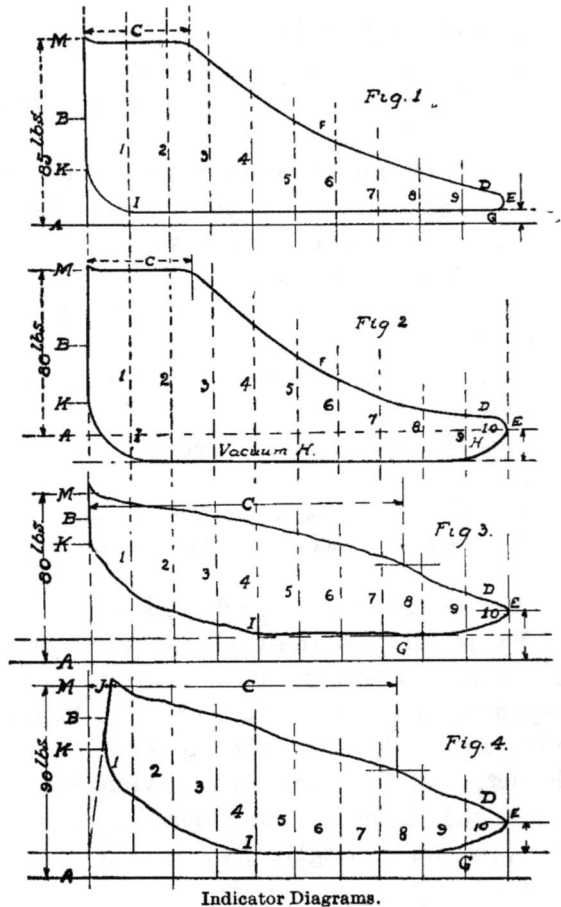

Indicator Diagrams.

The indicator diagram, Fig. 1, represents a card taken from a non-condensing automatic cut-off engine under 85 lbs. boiler pressure which cut-off at one quarter stroke; diagram Fig. 2 represents a card taken from a condensing, automatic cut-off engine under 80 lbs. boiler pressure which cut-off at one quarter stroke; by comparison of the two diagrams it will be seen that the lines are exactly the same with exception (G) back pressure, Fig. 1 and (H) Vacuum, Fig. 2 and indicates that the condensing engine has the advantage, in economy and power, over the non-condensing; everything else being equal.

Fig. 3 represents a card taken from a plain slide valve engine, as is usually used in farm engine practice, under 80 lbs. boiler pressure, and which cut-off at three quarter stroke.

Fig. 4 represents a card taken from the same style of an engine in which the valve had no lead, this is indicated at (J) and shows a loss of power as the piston traveled the first part of its stroke without the full pressure.

A comparison of Fig. 1 and Fig. 3 shows the advantage in economy of the automatic cut-off engine which cuts off at ¼ stroke over the plain slide valve engine which cuts off at ¾ stroke. This is shown in Fig. 1 by line (F) which is the expansion line.

In Figs. 1, 3 and 4 is shown, by the lower line (G), the back pressure in the cylinder or the pressure above atmospheric pressure; this is largely due to the necessity, in farm engine practice, of having the exhaust nozzled to produce the required draft for good combustion, while in Fig. 2 this line (H) is far below the atmospheric line and is due to the vacuum, produced by the condenser, which can be used only where the required draft is obtained by a chimney or other artificial means.

The comparison of these cards is to show the economy of a condensing over a high pressure engine and to give the reader an idea of the sensibilities and workings of the indicator, to show up the defects and efficiency of the different styles of engines as to economy and the power developed.

After an indicator card is taken it is divided into ten equal parts, as shown in diagrams, and the pressure represented between each section is figured mathematically to the scale furnished with every indicator to correspond to the spring used; then these amounts are added together and divided by ten, the result being the mean-effective pressure in the cylinder during one full stroke of the piston from which the power developed by the engine can be figured by rule given on page 211.

HYDRAULIC BOILER TEST.

In preparing to test a boiler, of any size or style, with cold water pressure, special attention must be paid to having the boiler full of water, and a reliable steam gauge.

To be sure that the boiler is full, remove the safety valve, whistle or some connection that will leave an opening in the highest point in the dome. This opening will let the air out as the water raises and will show when the boiler is full by the water overflowing, then replace the connection removed.

Never take it for granted that the boiler is full of water. You should know that it is by following the above instructions, as, should an attempt be made to pump the boiler full of water, with all fittings and connections air tight, a large quantity of air would be compressed in the dome and upper part of boiler, as it has no means of escape; now if the extra pressure necessary for the test be put upon the boiler, the compression of air within the boiler would be very liable to cause an explosion which would have as much force and be as disastrous as would be occasioned by an explosion from over pressure of steam; while if the boiler contained no air and burst under the hydraulic pressure there would be no explosion, simply a fracture at the weakest point, for the reason that the instant a small quantity of water escaped the pressure is relieved and no serious damage would occur.

After the boiler is filled with water see that all valves, fittings and outlets to the boiler, except the steam gauge, are closed and tight; then make the connection from the force pump to the most convenient pipe that leads into the boiler and the desired pressure can be raised by means of the force pump, and be registed by the steam gauge.

In testing a boiler its condition and the manner in which it has been handled should be considered, though in every case the pounds pressure of water to which it is subjected should be about double that of the amount of steam pressure to be carried, this will leave a safe working margin.

WHISTLE SIGNALS.

A regular code of signals should be adopted and used by every engineer no matter what kind of an engine or for what purpose used; then every one in connection with the business will soon become familiar with them and understand what they mean.

It should be the engineer's duty to sound the whistle and no one else should touch it unless authorized to do so by the engineer. The less the whistle is sounded the better, as then the sounding of the regular code of signals will attract the attention of all within hearing.

CODE OF SIGNALS FOR STATIONARY ENGINEERS.

One long sound at morning, noon and night to indicate the time to begin and stop work.

One short sound should be given about five minutes before the starting time to warn all hands that the machinery is about to start.

One long continuous sound or a long succession of sounds will indicate fire and is a call for help.

CODE OF SIGNALS FOR TRACTION ENGINEERS.

One long sound in the morning or at noon will indicate the working place.

Two short sounds means that the engine is about to start and work to begin.

One short sound means to stop.

Two long sounds means that the job or the day's work is completed.

Three medium short sounds means that the grain haulers should hurry as the machine is about to wait for them.

One moderately long sound followed by three short ones means that the water supply is getting low, and water hauler must hurry with fresh supply.

A rapid succession of short sounds means fire or distress, and should be promptly responded to by all within hearing.

Don't jerk the whistle valve open suddenly, it should be opened and closed gradually and the time of each sound and the pauses between them should be well timed and of equal length.

By screwing the bell of the whistle up or down will change the tone or pitch of the sound.

By screwing the bell down, a sharper and more piercing sound will be produced, by screwing the bell up the sound will be of a lower pitch and can be heard at a greater distance.

When the bell of whistle is set to a certain pitch secure it by check nut on top, then the sound will become familiar and can be distinguished from others.

RULES AND TABLES.

To find the steam pressure on slide valve, multiply the unbalanced area of valve in inches by pounds pressure of steam per square inch, add weight of valve in pounds, and multiply the sum by 0.15.

Safety boiler pressure according to the United States Government rule is as follows: Multiply ⅙ of the lowest tensile strength found on any plate in the cylindrical shell by the thickness expressed in inches or part of an inch of the thinnest plate in the same cylindrical shell, and divide by the radius or half the diameter, also expressed in inches, and the sum will be the pressure allowable per square inch of surface for single riveting, to which add 20 per centum for double riveting.

To find the water pressure on steam pipes leading from boiler to steam gauge, divide the difference in height between the highest point of pipe and the center of steam gauge by 2³⁄₁₀; the result will be the pressure exerted by the water in the pipe in pounds upon the gauge.

Area of a Circle.—To find the area of a circle when the diameter is given, multiply the diameter by itself, or in other words square the diameter and multiply the result by .7854. Ex. Diameter 5 inches, 5x5=25x.7854=19.635 area.

Circumference of a Circle.—To find the circumference of a circle when the diameter is given, multiply the diameter by 3.1416. Ex. Diameter is 5 inches. 5 x 3.1416 = 15.708 inches circumference.

Diameter of a Circle.—To find the diameter of a circle when the circumference is given, multiply the circumference by .31831. Ex. Circumference 20 inches. 20 x .31831 = 6.366 diameter.

To find the pressure of pounds per square inch of a column of water, multiply the height of the column in feet by .434. Approximately, we generally call every foot elevation equal to one-half pound pressure per inch, this allows for ordinary friction.

To find the diameter of a pump cylinder to move a given quantity of water per minute (100 feet of piston being the standard of speed), divide the number of gallons by 4, then extract the square root and the product will be the diameter in inches.

To find the horse power necessary to elevate water to a given height, multiply the total weight of water in pounds by the height in feet and divide the product by 33000 (an allowance of 25% should be added for friction, etc.).

The area of the steam piston multiplied by the steam pressure, gives the total amount of pressure that can be exerted. The area of the water piston multiplied by the pressure of water per square inch gives the resistance. A margin must be made between the power and the resistance to move the piston at the required speed—say 50%.

To find the capacity of a cylinder in gallons, multiply the area in inches by the height of stroke in inches, which gives the total number of cubic inches; divide this amount by 231 (which is the cubic contents of a gallon in inches), the product is the capacity in gallons.

RULES FOR CALCULATING THE SPEED OF GEARS AND PULLEYS.

In calculating for pulleys, multiply or divide by their diameter in inches.

In calculating for gears, multiply or divide by the number of teeth required.

The driving wheel is called the "driver," and the driven wheel the "driven."

PROBLEM 1.—To find the diameter of the driver when the revolutions of the driver and driven and the diameter of the driven are given.

RULE.—Multiply the diameter of driven by its number of revolutions, and divide by the number of revolutions of the driver.

PROBLEM 2.—To find the diameter of the driven to make the same number of revolutions in the same time as the driver, when the diameter and revolutions of driver are given.

RULE.—Multiply the diameter of the driver by its number of revolutions, and divide the product by the required number of revolutions.

PROBLEM 3.—To find the number of revolutions of the driven when the diameter or number of teeth and number of revolutions of the driver and the diameter or number of teeth of the driven are given.

RULE.—Multiply the diameter or number of teeth of the driver by its number of revolutions, and divide by the diameter or number of teeth of the driven.

PROBLEM 4.—To find the number of revolutions of the driver when the diameter of the driven and the number or revolutions of driven are given.

RULE.—Multiply the diameter of driven by its number of revolutions, and divide by the diameter of the driver.

SHAFTS AND PULLEYS.

To find the size of pulley needed to give the main shaft a certain number of revolutions, multiply the diameter of fly wheel in inches by the number of revolutions of the engine, and divide by revolutions of main shaft.

To find the revolutions of main shaft when diameter of its pulley is given, multiply the diameter of fly wheel in inches by its number of revolutions, and divide by the diameter of pulley in inches.

To find diameter of fly wheel needed to drive a pulley at a given number of revolutions, multiply the diameter of pulley in inches by its number of revolutions, and divide by number of revolutions of the engine.

To find revolutions of an engine with a given fly wheel to drive a pulley at a given number of revolutions, multiply the diameter of pulley in inches by its number of revolutions, and divide by the diameter of fly wheel in inches.

RULE FOR FINDING THE AMOUNT OF HEATING SURFACE IN A LOCOMOTIVE BOILER:

Find surface of flues by multiplying the diameter by 3.14 to get circumference, and multiply this product by the length of the flue, then multiply this result by the number of flues in the boiler, and divide by 144 to get number of square feet of surface in the flues.

Multiply length and width of fire box for number of square inches in crown sheet, then multiply the length and height of the fire box and the result by two, which gives the number of square inches in the sides; then multiply the width and height and multiply the result by two, which gives the number of square inches in the ends, from which subtract number of square inches in space left for door and flues —then add all these results together and divide by 144 to get the number of square feet in the fire box. Add same to number of square feet of flues, and the total will be the number of square feet in the boiler.

By dividing number of square feet by rated horse power of boiler will give number of square feet of heating surface to each horse power.

TABLES.

ALLOYS.

ALLOYS.	Tin.	Copper.	Zinc.	Antimony.	Lead.	Bismuth.
Brass, engine bearings.......	13	112	¼	....	....	....
Tough Brass, engine work...	15	100	15	....	....	....
Tough, for heavy bearings...	25	160	5	....	....	....
Yellow Brass, for turning...	....	2	1	....	....	....
Flanges to stand brazing....	....	32	1	....	1	....
Bell Metal	5	16	....	....	....	....
Babbitt's Metal	10	1	....	1	....	....
Brass Locomotive Bearings.	7	64	1	....	....	....
Brass, for straps and glands.	16	130	1	....	....	....
Muntz's Sheathing	....	6	4	....	....	....
Metal, to expand in cooling..	....	....	....	2	9	1
Pewter	100	....	....	17	....	....
Spelter	....	1	1	....	....	....
Statuary Bronze.............	2	90	5	....	2	....
Type Metal, from	....	....	....	1	3	....
Type Metal, to	....	....	....	1	7	....
SOLDERS.						
For Lead...................	1	....	....	....	1½	....
For Tin	1	....	....	....	2	....
For Pewter	2	....	....	....	1	....
For Brazing (hardest)	....	3	1	....	....	....
For Brazing (hard)..........	....	1	1	....	....	....
For Brazing (soft)..........	1	4	3	....	....	....
For Brazing (soft).........or	2	....	....	1	....	....

CIRCUMFERENCE AND AREAS OF CIRCLES.

Diameter.	Circumference.	Area.	Diameter.	Circumference.	Area.
1/32	.0981	.00076	3/4	11.78	11.044
1/16	.1963	.00306	4	12.56	12.566
1/8	.3926	.01227	1/4	13.35	14.186
3/16	.5890	.02761	1/2	14.13	14.904
1/4	.7854	.04908	3/4	14.92	17.720
5/16	.9817	.07669	5	15.90	19.635
3/8	1.178	.1104	1/4	17.49	21.647
7/16	1.374	.1503	1/2	17.27	23.758
1/2	1.570	.1963	3/4	18.06	25.967
9/16	1.767	.2485	6	18.84	28.274
5/8	1.963	.3067	1/4	19.63	30.679
11/16	2.159	.3712	1/2	20.42	33.183
3/4	2.356	.4417	3/4	21.20	35.784
13/16	2.552	.5184	7	21.99	38.484
7/8	2.748	.6013	1/4	22.77	41.282
15/16	2.945	.6902	1/2	23.56	44.178
1	3.141	.7854	3/4	24.34	47.173
	3.534	.9940			
1/4	3.927	1.227	8	25.13	50.265
	4.319	1.484	1/4	25.91	53.456
1/2	4.712	1.767	1/2	26.70	56.745
	5.105	2.073	3/4	27.48	60.132
3/4	5.497	2.405	9	28.27	63.617
	5.890	2.761	1/4	29.05	67.200
2	6.283	3.141	1/2	29.84	70.882
	6.675	3.546	3/4	30.63	74.662
1/4	7.068	3.976			
	7.461	4.430	10	31.41	78.539
1/2	7.854	4.908	1/4	32.20	82.516
	8.246	5.411	1/2	32.98	86.590
3/4	8.639	5.939	3/4	33.77	90.762
	9.032	7.491	11	34.55	95.033
3	9.424	7.068	1/4	35.34	99.402
1/4	10.21	8.295	1/2	36.12	103.86
1/2	10.99	9.621	3/4	36.91	108.43

Circumference and Areas of Circles—Continued.

Diameter.	Circumference.	Area.	Diameter.	Circumference.	Area.
12	37.69	113.09	21	65.97	346.36
¼	38.48	117.85	½	67.54	363.05
½	39.27	122.71	22	69.11	380.13
¾	40.05	127.67	½	70.68	397.60
13	40.84	132.73	23	72.25	415.47
¼	41.62	137.88	½	73.82	433.73
½	42.41	143.13	24	75.39	452.39
¾	43.19	148.48	½	76.96	471.43
14	43.98	153.93	25	78.54	490.87
¼	44.76	159.48	½	80.10	510.70
½	45.55	165.13	26	81.68	530.93
¾	46.33	170.87	½	83.25	551.54
15	47.12	176.78	27	84.82	572.55
¼	47.90	182.65	½	86.39	593.95
½	48.69	188.69	28	87.96	615.75
¾	49.48	194.82	½	89.53	637.94
16	50.26	201.06	29	91.10	660.52
¼	51.05	207.39	½	92.67	683.49
½	51.83	213.82	30	94.24	706.86
¾	52.62	220.35	½	95.81	730.61
17	53.40	226.98	31	97.38	754.76
¼	54.19	233.70	½	98.96	779.31
½	54.97	240.52	32	100.5	804.24
¾	55.76	247.45	½	102.1	829.57
18	56.54	254.46	33	103.6	855.30
¼	57.33	261.58	½	105.2	881.41
½	58.11	268.80	34	106.8	907.92
¾	58.90	276.11	½	108.3	934.82
19	59.69	283.52	35	109.9	962.11
¼	60.47	291.03	½	111.5	989.80
½	61.26	298.64	36	113.0	1017.8
¾	62.04	306.35	½	114.6	1046.3
20	62.83	314.16	37	116.2	1075.2
½	64.40	330.06	½	117.8	1104.4

Circumference and Areas of Circles—Continued.

Diameter.	Circumference.	Area.	Diameter.	Circumference.	Area.
38	119.3	1134.1	½	155.5	1924.4
½	120.9	1164.1	50	157.0	1963.5
39	122.5	1194.5	½	158.6	2002.9
½	124.0	1225.4	51	160.2	2042.8
40	125.6	1256.6	½	161.7	2083.0
½	127.2	1288.2	52	163.3	2123.7
41	128.8	1320.2	½	164.9	2164.7
½	130.3	1352.5	53	166.5	2206.1
42	131.9	1385.4	½	168.0	2248.0
½	133.5	1418.6	54	169.6	2290.2
43	135.0	1452.2	½	171.2	2332.8
½	136.0	1486.1	55	172.7	2375.8
44	138.2	1520.5	½	174.3	2419.2
½	139.8	1555.2	56	175.9	2463.0
45	141.3	1590.4	½	177.5	2507.1
½	142.9	1625.9	57	179.0	2551.7
46	144.5	1661.9	½	180.6	2566.7
½	146.0	1698.2	58	182.2	2642.0
47	147.6	1734.9	½	183.7	2687.8
½	149.2	1772.0	59	185.3	2733.9
48	150.7	1809.5	½	186.9	2780.5
½	152.3	1847.4	60	188.4	2827.4
49	153.9	1885.7	½	190.0	2874.7

EFFECTIVE PRESSURE OF STEAM ON PISTON.

With different rates of expansion, boiler pressure being assumed as 100 lbs. per square inch.

Steam cut off at ¾ of stroke = 90 lbs. effective pressure.
" " " ⅔ " " = 80 " " "
" " " ½ " " = 69 " " "
" " " ⅓ " " = 50 " " "
" " " ¼ " " = 40 " " "

MEASURE OF LENGTH.

12	inches	1 foot.
3	feet	1 yard.
2	yards	1 fathom.
16½	feet	1 rod.
4	rods	1 chain.
10	chains	1 furlong.
8	furlongs	1 mile.
3	miles	1 league.

MEASURE OF VOLUME.

A cubic foot has .... 1728 cubic inches.
An ale gallon has .... 282 " "
A standard or wine gallon has ... 231 " "
A dry gallon has .... 268 8 " "
A bushel has .... 2150 4 " "
A cord of wood has .... 128 " feet.
A perch of stone has .... 24 75 " "
A ton of round timber has .... 40 " "
A ton of hewn timber has .... 50 " "
A box 19⅜ x19⅜ ins., 19⅜ ins. deep, contains 1 barrel.
A " 12¹⁵⁄₁₆x12¹⁵⁄₁₆ " 12¹⁵⁄₁₆ " " " 1 bushel.
A " 8⅛ x 8⅛ " 8⅛ " " " 1 peck.
A " 6⁷⁄₁₆ x 6⁷⁄₁₆ " ·6⁷⁄₁₆ " " " ½ "
A " 4¹⁄₁₆ x 4¹⁄₁₆ " 4¹⁄₁₆ " " " 1 quart.
An acre contains .... 4840 sq. yds.
209 feet long by 209 feet broad is .... 1 acre.

LIQUID MEASURE.

A barrel holds	31½	gallons.
A hogshead holds	63	"
A tierce "	42	"
A puncheon "	84	"
A tun "	252	"

BARREL MEASURE IN WEIGHT.

A barrel of flour is	196	pounds.
A barrel of pork is	200	"
A barrel of rice is	600	"
A firkin of butter is	56	"
A tub of butter is	84	"

WEIGHT OF CAST IRON BALLS.

	Lbs.		Lbs.
2 inch diameter	1.09	5½ inch diameter	22.68
2½ " "	2.13	6 " "	29.48
3 " "	3.68	6½ " "	37.44
3½ " "	5.84	7 " "	46.76
4 " "	8.73	7½ " "	57.52
4½ " "	12.42	8 " "	69.81
5 " "	17.04		

WEIGHTS AND MEASURES.

AVOIRDUPOIS OR COMMERCIAL WEIGHT.

16 drachms	1 ounce.
16 ounces	1 pound.
14 pounds	1 stone.
28 pounds	1 quarter.
4 quarters	1 cwt.
2240 pounds	1 long ton.
2000 pounds	1 ton.

SQUARE MEASURE.

144	square	inches	1 square foot.
9	"	feet	1 " yard.
30¼	"	yards	1 " rod.
40	"	rods	1 " rood.
4	"	roods	1 " acre.
640	"	acres	1 " mile.

TABLE OF DISTANCE.

A mile is	5280 feet or 1760 yards.
A knot is	6086 feet.
A league is	3 miles.
A fathom is	6 feet.
A metre is	3 feet 3⅜ inches.
A hand is	4 inches.
A palm is	3 "
A span	9 "
A hair is equal to	1/48 of an inch.
A line is equal to	1/12 of an inch.

SHRINKAGE OF CASTINGS.

Cast Iron, ⅛ inch per lineal foot.
Brass, 3/16 inch per lineal foot.
Lead, ⅛ inch per lineal foot.
Tin, 1/12 inch per lineal foot.
Zinc, 5/16 inch per lineal foot.

WEIGHT OF ROUND AND SQUARE ROLLED IRON PER LINEAL FOOT.

Inch.	Round.	Square.	Inch.	Round.	Square.
¼	.165	.211	4½	53.760	68.448
⅜	.373	.475	4¾	59.900	76.264
½	.663	.845	5	66.350	84.480
⅝	1.043	1.320	5¼	73.172	93.168
¾	1.493	1.901	5½	80.304	102.24
⅞	2.032	2.588	5¾	87.776	111.75
1	2.654	3.380	6	95.552	121.66
1⅛	3.359	4.278	6¼	103.70	132.04
1¼	4.147	5.280	6½	112.16	142.81
1⅜	5.019	6.390	6¾	120.96	154.01
1½	5.972	7.604	7	130.04	165.63
1⅝	7.010	8.926	7¼	139.54	177.67
1¾	8.128	10.352	7½	149.32	190.13
1⅞	9.333	11.883	7¾	159.45	203.02
2	10.616	13.520	8	169.85	216.33
2⅛	11.988	15.263	8¼	180.69	230.06
2¼	13.440	17.112	8½	191.80	244.22
2⅜	14.975	19.066	8¾	203.26	258.8
2½	16.588	21.120	9	215.04	273.79
2⅝	18.293	23.292	9¼	227.15	289.22
2¾	20.076	25.56	9½	239.60	305.056
2⅞	21.944	27.939	9¾	252.37	321.33
3	23.888	30.416	10	265.40	337.92
3¼	28.040	35.704	10¼	278.92	355.30
3½	32.512	41.408	10½	292.68	372.70
3¾	37.332	47.534	10¾	306.80	390.80
4	42.464	54.084	11	321.21	409.00
4¼	47.952	61.055	12	382.20	486.70

TABLE OF THE CAPACITY OF CISTERNS IN GALLONS

For Each 10 Inches of Depth.

Diam. in Feet.	Gallons	Diam. in Feet.	Gallons.	Diam. in Feet.	Gallons.	Diam. in Feet.	Gallons.
2	19.5	5	122.4	8	313.33	12	705.0
2½	30.6	5½	148.10	8½	353.72	13	827.4
3	44.06	6	176.25	9	396.56	14	959.6
3½	59.97	6½	206.85	9½	461.4	15	1101.6
4	78.33	7	239.88	10	489.2	20	1958.4
4½	99.14	7½	275.4	11	592.4	25	3059.9

The American Standard gallon contains 231 cubic inches, or 8⅓ pounds of pure water. A cubic foot contains 62.3 pounds of water, or 7.48 gallons. Pressure per square inch is equal to the depth or head in feet multiplied by .433. Each 27.72 inches of depth gives a pressure of one pound to the square inch.

MELTING POINT OF METALS, ETC.

Names.	Fahr.	Names.	Fahr.
Platina	4590°	Wrought Iron	2900°
Antimony	842	Steel	2500
Bismuth	487	Copper	2000
Tin	475	Glass	2377
Lead	620	Beeswax	151
Zinc	700	Sulphur	239
Cast Iron	2100	Tallow	92
Gold	2192	Silver	1832

WEIGHT OF METALS PER CUBIC FOOT.

	Lbs.		Lbs.
Brass	525	Lead, cast	710
Copper	550	Silver	655
Gold	1210	Steel	490
Iron, Cast	450	Tin, cast	456
Iron, Wrought	485	Zinc	450

HORSE POWER LINE SHAFTING.

Will transmit with Safety, Bearings say 8 to 10 ft. centres.

Diam. of Shaft in Inches.	Horse Power in one Rev.	Diam. of Shaft in Inches.	Horse Power in one Rev.	Diam. of Shaft in Inches.	Horse Power in one Rev.
15/16	.008	2 15/16	.216	5 15/16	1.728
1 3/16	.0156	3 3/16	.272	6 7/16	2.195
1 7/16	.027	3 7/16	.343	6 15/16	2.744
1 11/16	.043	3 11/16	.424	7 7/16	3.368
1 15/16	.064	3 15/16	.512	7 15/16	4.096
2 3/16	.091	4 7/16	.728	8 7/16	4.912
2 7/16	.125	4 15/16	1.00	8 15/16	5.824
2 11/16	.166	5 7/16	1.328	9 7/16	6.848

For Jack Shafts, or main section of Line Shafts, allow only three-fourths of the horse power given above, and also provide extra bearings wherever heavy strains occur, as in main belts or gears.

HALF-ROUND, OVAL AND HALF-OVAL IRON.

Weight per Lineal Foot.

Size Half Round.	Size Oval.	Weight per foot.	Size Half Oval.	Weight per foot.
3/8	3/8 x 3/16	.186	3/8 x 3/32	.093
7/16	7/16 x 7/32	.253	7/16 x 7/64	.127
1/2	1/2 x 1/4	.331	1/2 x 1/8	.166
5/8	5/8 x 5/16	.517	5/8 x 5/32	.259
3/4	3/4 x 3/8	.744	3/4 x 3/16	.372
7/8	7/8 x 7/16	1.013	7/8 x 7/32	.507
1	1 x 1/2	1.323	1 x 1/4	.662
1 1/8	1 1/8 x 9/16	1.624	1 1/8 x 9/32	.812
1 1/4	1 1/4 x 5/8	2.067	1 1/4 x 5/16	1.034
1 1/2	1 1/2 x 3/4	2.976	1 1/2 x 3/8	1.488
1 3/4	1 3/4 x 7/8	4.050	1 3/4 x 7/16	2.026
2	2 x 1	5.290	2 x 1/2	2.645

WEIGHT OF FLAT ROLLED IRON, PER FOOT.

Breadth.	Thickness.	Weight.	Breadth.	Thickness.	Weight.	Breadth.	Thickness.	Weight.
1 in.	⅛	.422	1¾ in.	1⅛	6.653	2¼ in.	2	15.208
.....	¼	.845	.....	1¼	7.393	.....	2⅛	16.158
.....	⅜	1.267	.....	1⅜	8.132	2½ in.	⅛	1.056
.....	½	1.690	.....	1½	8.871	.....	¼	2.112
.....	⅝	2.112	.....	1⅝	9.610	.....	⅜	3.168
.....	¾	2.534	2 in.	⅛	.845	.....	½	4.224
.....	⅞	2.956	.....	¼	1.689	.....	⅝	5.280
1¼ in.	⅛	.528	.....	⅜	2.534	.....	¾	6.336
.....	¼	1.056	.....	½	3.379	.....	⅞	7.392
.....	⅜	1.584	.....	⅝	4.224	.....	1	8.448
.....	½	2.112	.....	¾	5.069	.....	1⅛	9.504
.....	⅝	2.640	.....	⅞	5.914	.....	1¼	10.560
.....	¾	3.168	.....	1	6.758	.....	1⅜	11.616
.....	⅞	3.696	.....	1⅛	7.604	.....	1½	12.672
.....	1	4.224	.....	1¼	8.448	.....	1⅝	13.728
.....	1⅛	4.752	.....	1⅜	9.294	.....	1¾	14.784
1½ in.	⅛	.633	.....	1½	10.138	.....	1⅞	15.840
.....	¼	1.266	.....	1⅝	10.983	.....	2	16.896
.....	⅜	1.900	.....	1¾	11.828	.....	2⅛	17.952
.....	½	2.535	.....	1⅞	12.673	.....	2¼	19.008
.....	⅝	3.168	2¼ in.	⅛	.950	.....	2⅜	20.064
.....	¾	3.802	.....	¼	1.900	2¾ in.	¼	2.323
.....	⅞	4.435	.....	⅜	2.851	.....	½	4.647
.....	1	5.069	.....	½	3.802	.....	¾	6.970
.....	1⅛	5.703	.....	⅝	4.752	.....	1	9.294
.....	1¼	6.337	.....	¾	5.703	.....	1¼	11.617
.....	1⅜	6.970	.....	⅞	6.653	.....	1½	13.940
1¾ in.	⅛	.739	.....	1	7.604	.....	1¾	16.264
.....	¼	1.479	.....	1⅛	8.554	.....	2	18.587
.....	⅜	2.218	.....	1¼	9.505	.....	2¼	20.910
.....	½	2.957	.....	1⅜	10.455	.....	2½	23.234
.....	⅝	3.696	.....	1½	11.406	3 in.	¼	2.535
.....	¾	4.435	.....	1⅝	12.356	.....	½	5.069
.....	⅞	5.178	.....	1¾	13.307	.....	¾	7.604
.....	1	5.914	.....	1⅞	14.257	.....	1	10.138

HORSE POWER BELTING.

Will transmit with Safety.

Width of Belt in Inches.	Horse Power per 100 feet Velocity of Belt.		Width of Belt in Inches.	Horse Power per 100 feet Velocity of Belt.	
	Single Belt.	D'ble Belt.		Single Belt.	D'ble Belt.
1	.09	.18	12	1.09	2.18
2	.18	.36	14	1.27	2.55
3	.27	.55	16	1.45	2.91
4	.36	.73	18	1.64	3.27
5	.45	.91	20	1.82	3.64
6	.55	1.09	22	2.00	4.00
7	.64	1.27	24	2.18	4.36
8	.73	1.46	28	2.55	5.09
9	.82	1.64	32	2.91	5.82
10	.91	1.82	36	3.27	6.55
11	1.00	2.00	40	3.64	7.27

In the calculations for horse power in the above table, the belt is assumed to run about horizontally, the semi-circumference of smaller pulley has been considered as the ordinary arc contact of belt. Any reduction of this contact will make approximate proportional reduction of horse power.

www.ingramcontent.com/pod-product-compliance
Lightning Source LLC
LaVergne TN
LVHW051546070426
835507LV00021B/2434